D0952553

The Immortal Cell

One Scientist's Quest to Solve the Mystery of Human Aging

MICHAEL D. WEST

DOUBLEDAY

New York   London   Toronto   Sydney   Auckland

Acknowledgments

This book represents an attempt to draw a limited synthesis between such disparate areas of human thought as ancient mythology and modern molecular biology. As a result, I have utilized far too many sources to give adequate credit here. Some excellent examples are highlighted in "Further Reading" and some others can be found in the "Notes" posted on my Web site, michaelwest.org.

I am also indebted to many colleagues for affording me the opportunity to find a voice for many of the concepts discussed in this book. In particular, I wish to acknowledge Keith H.M.S. Campbell, Leonard Hayflick, Robert P. Lanza, Mary Ann Liebert, and Woodring E. Wright for helpful discussions.

I also wish to thank Peter Israel for his excellent editorial advice and Roger Scholl and Chris Fortunato for their patient and critical editorial assistance.

Lastly, I wish to thank my wife, Karen, for her many hours of patient support.

PUBLISHED BY DOUBLEDAY
a division of Random House, Inc.

DOUBLEDAY and the portrayal of an anchor with a dolphin are
registered trademarks of Random House, Inc.

Book design by Erin L. Matherne and Tina Thompson

Cataloging-in-Publication Data is on file with the Library of Congress.

ISBN 0-385-50928-6

PRINTED IN THE UNITED STATES OF AMERICA

[October 2003]

FIRST EDITION

1 2 3 4 5 6 7 8 9 10

Contents

One spring weekend during my freshman year of college, I turned my back on class work to hitchhike from Troy, New York, to Bennington, Vermont. My goal was simply to get a sense of life in New England. The driver dropped me off just short of the goal, near a small cemetery at the outskirts of the town. As I waited for my next ride, I wandered among the graves, reading about the lives of those who had died. I happened upon the tombstone of a young girl. There, beneath her name and a number range indicating the years she lived, were the immortal words of Shakespeare taken from *King Richard:*

> *Of comfort no man speak:*
> *Let's talk of graves, of worms, and epitaphs;*
> *Make dust our paper, and with rainy eyes*
> *Write sorrow on the bosom of the earth.*

I was deeply touched by her parents' statement of grief. It reflected my deep-seated belief that the greatest of all human values is our love for one another. What is more important to us than the warmth of our lover's kiss, enduring friendships, the companionship of our family? But all of these people we know and love walk with us upon this same planet, a celestial sphere careening through a universe that operates under the cold and impersonal laws of nature. We can be certain that the sun will rise tomorrow at the exact place on the horizon and at the precise time, within a millionth of a second of the time predicted by the astronomers. Our hopes or aspirations cannot change that fact. Indeed, the sun will rise tomorrow and the day after that, as it has for millennia. But there is a stark reality where the warmth of human love conflicts with the coldness of nature. Just as certain as our predictions of the rotation of the earth, we can be sure that the sun will rise on a day when we and all of our loved ones will walk the earth no more.

Human aging and death is inarguably the greatest challenge ever faced by mankind. All of us grow old. We are all going to die.

The 1934 novel *Appointment in Samarra,* written by John O'Hara, opens with a short story attributed to Somerset Maugham about the servant of a businessman who went out shopping one day in Baghdad. Turning from a merchant, he found himself looking directly into the icy black eyes of Death himself, who shot him a terrifying and fixed stare. The servant turned tail and ran like there was no tomorrow back to the home of his master, begging him for a horse. The kindhearted master agreed, and the servant escaped to seek refuge in the neighboring city of Samarra. Later that day, when the businessman traveled to Baghdad to complete the shopping, he, too, encountered the personage of Death. He accosted Death and reprimanded him for the unnecessary cruelty of frightening his poor servant with his horrifying stare. Death replied, "It was only a start of surprise. I was astonished to see him in Baghdad, for I had an appointment with him tonight in Samarra."

Most of us deal with the fate of our earthly bodies in the same manner as the servant of Baghdad. We run. Our fear of death forces us to retreat into denial. And our denial is effective. While we readily acknowledge that people do age and die, we live as though it will never happen to us. Our strategy is to lose ourselves in the crowd, to busy ourselves in the details of everyday life. After all, Death will never find us; we are hiding in Samarra.

Old age is a predictable part of the cycle of life, just as certain as childhood, puberty, and adulthood. But its outcome is far less hopeful. Many of us will experience old age as described by an aged woman in the book *Aging in America:*

What is the sound I make when I am old? Shuffling for sustenance, napping for strength, dressing for no one, waiting for one certain visitor, I am a diminished me. No capital I. No self to fling free, a bird sailing skyward. There is only a small i, shriveled within the layers of the years. Somewhere in the early sleepless morning, when daylight still brings a flicker of promise, I lie young on my scarcely wrinkled bed and am warmed by the feel of husband-hands caressing me knowingly or of child-fingers on my face or of friend-touch on my hands. But these moments pass. Daytime brings no warmth, and at last I rise because I have always risen and go to prepare myself for a day which stays too shortly and a night which comes too soon. I am an island, barren, surrounded by the waters of my plight.

We all dread the inevitable signs of age, the first time we find a gray hair, the unwelcome feeling of arthritis in our joints, the crushing pain of a heart attack. But I think the reason death lies so heavy on the human heart is more profound than just this. In the course of our short lives, we grow to love one another. If we could live out our lives and never care for another human being, we would think nothing of the loss of a wife, a mother or father, sister or brother. We could view death as we do the falling of the leaves in autumn, bittersweet, perhaps, that summer is over, but understanding that the cycle of the seasons must march on, so there can be a new spring, and life will begin again. But I believe most of us love people very differently from the way we love a tree.

Our lover's cheek, the warm embrace of a mother greeting her daughter, the bright eyes of a child opening a birthday present, all these things we value as human beings, all these things that we cherish more than anything else in the world, are composed of flesh and blood, skin and bones, lymph and brains. The lesson from science couldn't be clearer on this point. The people we love are, from a scientific perspective, biological machines— machines not made of motors and gears, but of proteins, lipids, and DNA. And, like the machines we use every day, the human machine can and will break down. When a radio breaks, we throw it away. But when our husband or wife suffers an inoperable brain tumor, or our child contracts scarlet fever, we sense the grave implications of disease, the possibility of losing it all. The more we love, the more we chance to lose.

Medicine, operating on the assumption that the human body is ultimately a biological machine, has made measurable progress in the treatment of disease. There are more happy outcomes today than in times past. However, this advance has fallen short of its ultimate goal. No matter how well we treat the human machine, our flesh and blood are still mortal. No matter how much we may love someone, and wish life to go on happily ever after, no matter our wealth and the resources we can muster, no matter our social standing, no matter how hard we pray to heavenly deities, we stand powerless against the death of the ones we hold dearest. At some point they will never again see a Christmas tree, the intricate beauty of a snowflake, hear their favorite music, or hold our hand.

I know of no greater tragedy than this.

But while the body may decline in function slowly, like the gentle falling of a leaf in autumn, the human soul never ages. As

Victor Hugo described it: "The heart does not grow old, but it is said to dwell among ruins."

Our lives and the lives of our families are somewhat like a burning tenement house. We sit by the windows of our eyes, watching our bodies inevitably burn around us. Flesh and blood, cells and tissues, friends and lovers, age and die. The human body is engineered to flourish and then steadily degenerate, and that is that.

Or is it?

In the pages that follow, I tell the story of my own scientific journey to understand the meaning of life and death. It begins more traditionally, as a decade-long quest to find meaning within the pages of scripture. But it leads later to the brave new world of science and technology, to the discovery of the cellular "clock" and the mechanism that controls cell aging. Ultimately, it leads to the exploration of such controversial technologies as human therapeutic cloning and human embryonic stem cells, in an effort to turn back the biological aging process and regenerate organs and tissue. It is a story of a tempest of controversy as the potential of science and medicine confronts the deep-seated concerns of religion in the twenty-first century.

My motivation in writing this book is simple. Like the prophet Jeremiah, "the word is in my heart like a fire, a fire shut up in my bones. I am weary of holding it in." In the same way, I cannot rest until I tell this story of the new and better world science and medicine are creating. I believe this book is, in itself, a mere preface to a much larger story still in the making. For I feel confident that this will be the century when mankind finally begins to understand the biological basis of its own mortality and the mechanisms that may allow us to reverse, to an extent, the aging process. Like the gods of ancient folklore whose possession

of the secrets of mortality and immortality led to as much mis-
chief as acts of kindness, our possession of this knowledge will
be the truest test of our character. Will we use these technologies
to lift the burden of suffering from our fellow human beings, or
will we use them for our own selfish ends?

The ultimate goal of medicine will always be directed to alle-
viating the pain of human suffering. My hope is that the new science described in this book will be used to wipe the tears from
the eyes of the suffering, and to write joy "on the bosom of the
earth."

Chapter One

The Experiment of a Lifetime

How can a man be born when he is old? Surely he cannot enter a second time into his mother's womb to be born again?
—NICODEMUS (John 3:4)

Immortal life, life free of the debility of disease and death, has always been the dream of mankind. From the symbolic art decorating the tombs of ancient Egypt to the modern graveside eulogy, the quest for immortality is the very heart of what it is to be human. Yet in these opening years of the twenty-first century, that dream is far from being realized. You and I are still made of mortal cells. Every skin cell, every blood cell coursing through our veins, is programmed to age with time—and that is that. We are destined to grow old. As much as we may try to deny it, the sunny days of our youth will one day fade to the sunset of our old age. Like hostages, we are forced to watch the years slowly etch wrinkles into the brows of our loved ones, slip into our homes and rob us one by one of mother and father, husband and wife, brother and sister.

Twenty years ago, I set out as a young gerontologist with the goal of understanding this mysterious process. I knew from the start that the work would be challenging, that trying to find a means of intervening in its inexorable course would be like trying to invent a perpetual-motion machine, or reverse the rotation of the Milky Way galaxy. Aging, after all, has been one of the last mysteries to come into focus under the scientist's microscope. But though I expected the research to be difficult, I had no idea as a young scientist that my quest would lead me into international controversy, into a technology that would bring upon my colleagues and me the condemnation of both a president and a pope.

I was blissfully ignorant of the storm of controversy we would ignite one October evening in 2001 when my colleagues crowded into a ten-by-thirteen-foot windowless lab, which felt all the smaller because of all the equipment we'd packed into it. The room housed everything we needed to clone embryos: a lab bench, microscopes, a refrigerator, a specialized hood to keep materials sterile, and a cell culture incubator. We kept the room's door locked, under twenty-four-hour video surveillance, and marked on the outside with an orange BIOHAZARD logo.

The experiment we were to perform that night had been built upon twenty years of effort. Nevertheless, it was strikingly simple in design. On the table before us was a sophisticated microscope with black joysticks on either side. Each controlled the minute motion of microscopic glass needles, finely drawn shining glass filaments about the diameter of a human hair. On the left side of the microscope was a dish containing the "time machines"—that is, human egg cells. On the right were aged human body cells well into their allotted life span. All of the technology was finally in place when, like the time travelers in H. G. Wells's story, we pushed forward the joystick of a micro-manipulator. The significance of the exact moment set our hearts racing, for this experiment was designed to reverse the aging process of a cell by cloning a human embryo.

Earlier in the day, a reporter named Joannie Fischer with *U.S. News & World Report* had asked me, "Just what does it mean to play God?" and my knee-jerk answer had been, "Well, you know, every day when practitioners of the art of medicine administer antibiotics to a child dying of pneumonia, or give heart bypass surgery to someone dying of coronary disease, we play God."

But now, in the lab, I wondered if what I'd said to Joannie Fischer really was a complete answer. I knew very well that we were working with a powerful technology that could change the face of life as we knew it. *Were* we playing God? Were we trespassing in a realm in which we humans have no right to go? Where were the boundaries that separated what we *can* do from what we *should* do? Should the only restrictions on our behavior be those of the limits of our own imagination?

Then—as now—there was only one position that made sense to me. Every human being should have not only the right but the passionate duty to reach out with all his or her strength to help others, even if it involves such controversial technology as cloning. If that means playing God, then it is playing God in a good way. I realized that I would do anything to save a human life, short of harming a fellow human.

Cloning immediately brings to mind Aldous Huxley's *Brave New World*, flying-saucer cults, and images of Frankenstein and Adolf Hitler. It triggers ethicists to decry the entire enterprise of biotechnology as something that will unravel the very fabric of society. But to cell biologists and medical researchers, cloning is a doorway to freeing patients from the prison of the hospital bed, an opportunity to create new lifesaving technologies thought impossible only ten years ago. Today cloning is a two-edged sword. When used to clone a human being, what is often called *human reproductive cloning*, it could lead to a hubristic effort to create an identical twin, a feat that could carry a high risk of disaster for both child and mother. But our dream is of a different kind of cloning, what we call *human therapeutic cloning*, in order to find a way to reverse the aging of a human cell. We want to hijack the cloning process to perform the magic of age reversal, but then to jump off that track before the development of a baby begins. We believe that in taking an old cell back to the earliest stages of the embryo, we may be able to open the door to new strategies designed to inweave new young cells and tissues to meet the needs of millions of expectant patients.

This hopeful vision to replace the old and broken cells in our body with pristine healthy ones is called *regenerative medicine*. It is an enduring dream, one as old as human history. It is also, I believe, our destiny.

Our one-horse biotech company, Advanced Cell Technology (ACT for short), is located in Worcester, Massachusetts. Virtually every other weekend in 2001, several of us drove the forty-two miles from our offices in Worcester into Somerville, not far from Boston, with a portable incubator to collect a small harvest of human oocytes—or egg cells—culled from volunteers at a research clinic. The clinic is run by Ann Kiessling, a collaborator in the experiments, who is Director of the Bedford Research Foundation.

Our scientists then rushed the eggs back to our lab in Worcester, where, in the silken repose of a weekday evening, we performed the delicate transfer procedure, removing the nucleus from the eggs and transferring the nucleus from cells cultured from a human body.

But up to this point, to our enormous impatience, despite our spectacular success in obtaining embryos with nonhuman animals, our progress with human cells has been interminably slow.

The problem has not been one of lack of technique or expertise. We've learned lessons from the cloning of many dozens of cattle, and from experiments in pig, mouse, goat, and even endangered-animal cloning. All of these lessons were digested and absorbed in the hope that they would lead us to success in the human experiment we're performing this October night. But we haven't had access to enough egg cells.

Every day FedEx shipments of hundreds of healthy cow oocytes have arrived at our doorstep. They've cost, at most, a dollar apiece. The supply has been abundant, virtually unlimited, and as a result, our animal nuclear transfer program has progressed by leaps and bounds.

Human oocytes, however, have been far more difficult to find.

The human egg cell certainly takes the prize for being the most controversial cell in the human body. The bellicosity over federal funding for human embryo research has resulted in a "see no evil, fund no evil" hands-off policy in the government. As a result, there has been no federal grant support available for human embryo research. Instead, the work has taken place in private companies, and in the adverse economic climate of today we have been obliged to scrounge for the financial resources to obtain the necessary eggs.

They've come to us in small precious quantities, two here, four there. At the insistence of our Ethics Advisory Board, an independent group assembled to help us navigate the sometimes turbulent waters of controversy, the donor women who have participated in our program have been carefully screened, given rigorous physical examinations, and tested psychologically within an inch of their psyches, all of which have anted up the costs.

We have no shortage of potential donors. But each cycle, because of the extraordinary precautions we take, has cost us approximately $20,000, no small change for a fledgling company with one eye on next week's payroll and the other on the many millions of dollars we will need one day to actually deliver cures for human disease.

Now, on this cold and cloudy evening, October 9, 2001, Jose Cibelli, our vice president of research, had driven the hour-long ride back to our warm laboratory in the biotech complex abutting the University of Massachusetts Medical Center, on the edge of downtown Worcester. On the seat next to him was a small portable incubator that could easily be mistaken for a thermos bottle if it weren't for the fact that it was plugged into the cigarette lighter. A line of red LEDs assured Jose that the contents were warmed to 37 degrees centigrade, the normal temperature of the human body. Inside were eleven freshly donated human egg cells. In the front seat with Jose was a bodyguard, hired to ensure that every egg cell was accounted for and not used in some clandestine effort to clone a human being.

It was 7:00 P.M. when they reached our biotech complex. The parking lot was nearly empty. Only a few laboratory workers were unlocking their cars to head home after a long day at the bench. Our work was only beginning. Jose and the bodyguard slid the access card at the door and hurried across the polished floors of our lobby, bearing their precious cargo. Our offices were deserted too, the lights turned off, the lab rooms empty—except for the small, windowless human-cloning lab where our team was anxiously waiting.

Everyone felt the urgency of the experiment that night. Two months earlier, the U.S. House of Representatives had passed by a 100-vote margin the Human Cloning Prohibition Act of 2001, a bill that, if signed into law, would criminalize the very work we do, imposing on scientists penalties of ten years in prison and a million-dollar fine. The bill was passed despite the loud objections of such diverse groups as the Juvenile Diabetes Research Foundation International, the Alliance for Aging Research, the American Liver Foundation, the American Society of Cell Biology, and the Association of American Medical Colleges. Representative James Greenwood of Pennsylvania led an effort to pass a bill banning only the cloning of people, leaving human therapeutic cloning legal, but his words of caution were drowned out by cries of "industrial exploitation of human life" and "cloned human embryo farms." Later, Representative Dennis Kucinich of Ohio reminded the Congress, "The Creator that our founders referred to was not ACT." And the President himself asked for a ban on all uses of cloning, both in cloning a child and its use in medicine. Our only hope was that we could quickly demonstrate to the world how important this work is by showing how we could use it to cure disease. With precarious finances, who knew how many more experiments we could conduct? This was our seventh attempt.

With this in mind, we invited two of the most skilled nuclear transfer scientists available: a young scientist I'll refer to as C. K., and Teru Wakayama, a Japanese scientist best known for being the first to clone a mouse. Teru succeeded in cloning the mouse where others failed because of his work in pioneering some new techniques to inject a body cell directly into an egg cell.

The tasks of preparing the microtools were already completed. The pipettes, those delicate, precision hollow glass tubes that we use to handle the cells, had been cut, ground, and polished. One would be used, with a slight bit of suction, to hold the spherical egg cell in place under the microscope lens. Although the egg cell has the largest overall dimensions of any cell in the body, it is still microscopic, about 0.005 inches in diameter, smaller than the period at the end of this sentence. The other glass tube, much smaller in caliber than the one that holds the egg cell, would be used to pierce the leathery shell around the egg, called the *zona pellucida*, and remove its chromosomes. We used micromanipulation dishes with small round troughs in them, which allowed us to segregate groups of oocytes and use different media on them—different dyes, for example. A television screen overhead permitted the rest to watch the manipulation of the microscopic cells, magnifying not only the cells themselves but the pipettes C. K. maneuvered under the microscope lens.

During her reproductive years, a woman's ovaries release on average one mature egg cell ready to fuse with a sperm cell and make an embryo every month. We and others have found that these mature oocytes are the best egg cells to use for cloning. Our donors have agreed to receive a series of hormone shots that will induce the release of many more mature egg cells than normal.

This week's collection was eleven eggs. At least eight of them were mature. Sometimes none have been usable. Our hormone stimulation of the women donors was designed to be very safe, and therefore so gentle that we obtained far fewer egg cells than is typical in normal in vitro fertilization (IVF) procedures.

With a deep collective breath, the team now went to work. At 10:00 P.M. C. K. began by enucleating the first oocyte—a simple but highly delicate process. First the egg was placed in a culture medium under a thin layer of oil and soaked in a dye that adheres to DNA so that, when the egg is passed quickly under ultraviolet light, the egg's chromosomes glow a brilliant blue. The egg was held lightly in place by the pipette C. K. wielded in her left hand. She manipulated the other, smaller pipette with a joystick, which was hooked up to a microinjector that could be used either to remove or introduce material into the egg cell.

C. K. twirled the egg this way and that under the microscope, searching for the ideal spot to puncture it, an area closest to the chromosomes. Meanwhile, with a quick push of a foot pedal, she flooded the egg cell with ultraviolet light. The egg itself floated free inside the zona pellucida, which on the television screen looked like one of Saturn's rings. Under the ultraviolet light, the chromosomes shone at us like blue headlights through the foggy depths of the egg cell.

When she found the right spot, she anchored the egg firmly in position with the holding pipette. Then, her right hand on the joystick, she pushed the smaller pipette against the zona.

This is the key moment of enucleation, when things can go wrong. The room was deathly quiet except for the gentle humming of the electrical equipment and the *tick, tick, tick* of an inexpensive black plastic-rimmed clock hanging on the wall next to the microscope.

She was in a sense performing surgery on a single cell; in this case, instead of removing a diseased tonsil or gallbladder, she was doing a "chromosomectomy"—that is, removing the chromosomes containing all the DNA while doing as little damage as possible to the rest of the cell. It was with a surgeon's sure hand that C. K. did her meticulous work while the rest of us stared at the monitor. Clicking on a floor pedal activated an electronic device that set the small glass pipette vibrating very rapidly. This technique, pioneered by Teru, allowed her to turn what would normally be a rubbery and difficult-to-pierce membrane into something close to a piece of butter, the pipette a hot knife. She tunneled out a cylindrical hole to get access to the chromosomes. Then, with a light suction, she pulled the part of the egg cell containing the chromosomes into her pipette, swallowing the blob of chromosomes in a gulp. She then pulled the chromosomes away.

"I'm still getting the stringy cytoplasm," she complained.

Unlike that of any other species studied to date, the human enucleation left a bizarre trail of cytoplasm like warm mozzarella cheese.

Eventually the trail of cytoplasm snapped back. We all glanced at one another nervously. We hadn't seen that before in any other species we had cloned.

Done.

We all exhaled.

One down, seven to go. C. K. now confidently went after the other seven oocytes. One after another, she removed the genetic blueprint. We now had eight mature human oocytes prepared to receive a new blueprint and form an embryo.

Within fifteen minutes she finished her work. Each egg was returned to its culture medium in the warm and dark incubator to convalesce from the surgery. C. K. herself took a brief break, standing, stretching, chatting with the rest of us.

Watching her, I was reminded of the realities of medical research—the headaches, the strained eyes, the back muscle cramps from long hours at the lab bench, the afterimages of myriads of plastic tubes that linger in your eyes as you try to go to sleep. I thought of the disappointment of so many nights and weekends sacrificed for experiments that end up a failure. An old joke came to my mind: "You know why they call it research, don't you? Because you search and then you have to search again." But almost like a mother who is capable of suspending and suppressing the remembered pain of childbirth in order to go through pregnancy yet one more time, we keep returning to the lab bench one more weekend. It is the passion of discovery that brings us back, the thrill of glimpsing the tantalizing mysteries of nature, and that uncanny sense that comes over one peering fixedly through the microscope that somehow, right before his or her eyes, the future of medicine may be lurking.

At 11:00 P.M. Teru took over C. K.'s place at the bench. Teru is a gifted cloning scientist with "magic" hands. He quickly picked up one of the aged body cells and, clicking on the foot pedal, activated the vibrating needle to embed it deep within the egg cell.

Teru's innovation in cloning is largely the way he introduces the body cell. Before Teru, scientists would often place the body cell into the narrow pocket between the zona and the egg cell itself and then fuse them with a jolt of electricity. And that is where the references to Frankenstein likely originate. (In Mary Shelley's 1831 novel *Frankenstein*, the monster Dr. Frankenstein has created is resurrected from the dead with a vital jolt of electricity.) In our case, Teru did not use electricity at all, and the cells we were working with were very much alive all along.

"This one good!" he said in broken English, nodding his head.

Then he rolled his chair back to deal with the next one.

At 11:30 P.M. he handed the dish over to Jose.

We now had seven nuclear transfer units (NTs) that made it this far. But they would likely fail to develop on their own. They had the material of an egg cell and a human genetic blueprint, but were lacking one thing: the signal that a sperm cell normally imparts that sets the developmental ball rolling. Without activation the NTs would likely just sit there and slowly degenerate over the next few days. Beginning at 11:30 P.M., Jose transferred the seven NTs into a series of chemicals designed to simulate the activation signal normally given by the sperm cell. At 3:15 A.M. Jose retrieved the NTs and placed them in a rich medium designed to allow human embryos to grow. Sliding the small plastic dish with its precious cargo onto the stainless-steel shelf of the incubator, he then (at the request of our Ethics Board) dutifully padlocked it.

There, in that dark, warm nurturing environment, the real magic would occur. Within those egg cells millions of minute molecules were already busily at work repairing the damage inflicted by our crude surgery, and—with our fingers crossed—they would cast their rejuvenating spell over the old cell's DNA.

The body—or somatic—cell differs in at least one important respect from the sperm cell that normally enters the egg. Body cells normally have no chance of replicating forever in our children, and then our children's children, and so on. Only the reproductive lineage of cells has the potential to defy the inevitability of aging and death. Instead, our body cells age. They are mortal.

My hope was that the DNA from the aged body cells would be metamorphosed in the coming hours, transformed by the molecular machinery of immortality in the egg cell, taken back as if it were transported in a microscopic time machine. Of course, such hopes are generally dashed on the hard rocks of empirical science. But if we could accomplish this "miracle" of cell biology, it would be the first time in history that the aging of a human cell was actually reversed. And the implications would be profound. If we could do that for one cell, it naturally followed that we could then find a way to multiply the resulting young cells into millions of young cells that could be transplanted into the patient to repair and perhaps rejuvenate old tissues. This simple experiment using a few cells could one day become a means of treating millions of people suffering the debilities of old age.

The truth is that we'd been through this too many times to have unrealistic hopes, especially with a mere seven embryos. But still, the constantly moving mirage that the *next* experiment will yield exciting results is what motivates scientists. Like a habitual gambler, bit by bit we invest our lives and wait for the turn of the cards.

Finally, though, there was nothing left to do but to clean up, turn off the lights, and drive the deserted streets of Worcester, Massachusetts, back to our homes for a short night's sleep. We would dream of what the world could be, while the molecules churned within the reconstructed embryos, in that warm, dark incubator, in that small lab that may contain our future.

We wouldn't know anything for six days.

Chapter Two

The Green Face
of Osiris

The idea of immortality that like a sea has ebbed and flowed
in the human heart, with its countless waves of hope and fear,
beating against the shores and rocks of time and fate,
was not born of any book, nor of any creed, nor of any religion.
It was born of human affection, and it will continue to ebb
and flow beneath the mists and clouds of doubt and
darkness as long as love kisses the lips of death.

—ROBERT G. INGERSOLL
The Ghosts and Other Lectures, 1878

It was the summer of 1960. I was seven years old, bored to death by a marathon of baseball and war games. Hunting for something new, I led my friend Perry Johnson into our dark Michigan basement, where we looked for chemicals to "cook up" something. There in the dark passageway my father used to store old cans, I found an old rusty tin of pine turpentine, a can of shellac, carbon tetrachloride (then used as a dry-cleaning agent), and some charcoal from the nearby coal room. We lined them up on the concrete floor and began mixing them in a gallon glass jug. They bubbled and smelled. Intoxicated by the fumes, I poured out my concoction onto the floor. A black, gooey, flubberlike mass emerged.

What was this strange new substance? Had I invented something new? And most important of all: What else could I make?

That night I dreamed of chemistry and woke up obsessed.

Within days I had taken over the small coal room and made it a laboratory. I nailed together some old boards against one wall to make shelves for my chemicals and inventions. Until now, I had never been inclined to accept my father's offer to make a dollar an hour sweeping the greasy floors at his International Harvester truck dealership. But now I leapt at the opportunity and invested every dollar earned in Erlenmeyer flasks, condensers, and glass tubing. Every week a mail-order package chock-full of exotic chemicals arrived that in today's world would probably have brought the bomb squad knocking at our door.

Soon my coal-room laboratory was bursting at the seams. I had scavenged every discarded table and shelf as workspace, and the contents were spilling over into other rooms. My mother was increasingly concerned about the number of bottles labeled "poison" and "acid." Over dinner one night, I laid out my proposed resolution. I could move the lab equipment to the attic over our garage. "It would make a first-class laboratory," I argued. It would also reduce the risk of a fire that could burn down the whole house. (My parents knew about the explosion with my bromine experiment, but not about the one in the school storeroom.) I knew one more trip to the emergency room for stitches, or one more gas smelling up the house, would probably be the end of my obsession.

My father lobbied hard on my side. "We really aren't using the attic of the garage anyway," he pointed out, and he promised to watch over me to make sure I was safe. Between the two of us, we won the day. I had my new and spacious laboratory, and none too soon. My interests had by then expanded to physics, electronics, biology, and to chemistry experiments more volatile than coal and turpentine. I had discovered the joys of yellow phosphorus and nitroglycerine, materials far too dangerous for the all-American home.

This was the early 1960s, the age of *Sputnik* and the race for the moon. Science seemed to me to be a powerful, sometimes explosive force. Knowledge of the indivisibly and invisibly small could unleash unimaginably massive nuclear energy, even thermonuclear bombs. And similarly, it could unleash powerful insights that uncovered the mysteries of the origins of our universe. Gigantic telescopes could peer into the depths of space and observe events that occurred billions of years ago. Science offered the hope of curing disease and prolonging the life of people we knew and cared for. I loved science. I saturated myself in it, wore it the way one would wear a suit of clothes. I lived in my lab, emerging only for school and to sleep. Even from my bed, I strung wires from under my pillow, under the carpet, and through the heating ducts of the house into the lab to chart my brain waves during sleep. In those nights, when the first satellites circled overhead, I dreamed of devoting my life to technology, of using my talents to make a new and better world. I thought there was no limit to what we could accomplish through the power of science.

On Saturday, September 4, 1965, I sat in a camper on the back of a pickup truck along with my father and his friends in the White Cloud area of Michigan, deep in the woods by a small lake. On this night, a steady and heavy rain kept us huddled inside. We hovered around a small kitchen table, where we'd just finished eating the fish we had caught that day.

I went nowhere without bringing my science with me, and had planned for weeks on how to pack up a portable chemistry lab that now sat perched on top of a small refrigerator. While my father and the other fishermen drank beer and told tall tales, I sat quietly studying the specimens of "seaweed" I had collected earlier while snorkeling along a beautiful bank of weeds and swirling fish. I peered into a small microscope, enthralled with the beautiful green cells of the seaweed leaves. I furtively glanced at a Bunsen burner boiling the seaweed in a glass beaker to extract its "essence" in a dark green bubbling cauldron. The patter of the rain and the loud conversation over beer and cards at the table inches away nearly drowned out the radio in the background, but then I distinctly heard the name "Albert Schweitzer." I knew who Schweitzer was. I considered him to be one of the greatest men of medicine. I pressed my ear against the speaker. "Albert Schweitzer, the humanitarian, scientist, and theologian, known for his 'reverence for life,' is dead," the announcer said with a hollow voice.

I will always remember that moment with a profound sense of sadness, the sound of rain, the smell of fried fish and propane, and the sight of green liquid in my boiling beaker. "A reverence for life," I repeated to myself—"a reverence for life." I felt as if I were reciting a mantra. What was this strange feeling that had come over me? Somewhere nearby—in the communion of the death of Albert Schweitzer and the bubbling green essence before my eyes—I sensed some great truth, but only like a blind man knowing he is close to his dear home yet uncertain of where he stands or in precisely which direction it lies. I stared at the beaker. Something in my mind was directing me to the green in that vessel. I knew the green was chlorophyll, a pigment that extracts energy from sunlight. But no, my brain was pointing to something deeper, the symbolism of the green, not just the cold science of chlorophyll—the beauty and the mystery of life, a reverence for life symbolized by this color. This impression was to leave a permanent mark on my memory: the pull of curiosity to understand the central mystery of the meaning of life, and the pathos of the struggle of science and reason over suffering and death. *There is a green "essence" of life,* I realized, *and no one has yet captured it in a test tube.*

My father had been a bomber pilot in World War II and had settled down afterward as a mechanic and salesman rolled into one, a very hardworking and successful one at that. But he'd never finished high school. Neither had my mother. Doubtless as a result, and wanting a better life for me, they encouraged me in my pursuit of science. While they had no science education, I cannot think of a more supportive climate in which to nourish a budding scientist. Then, when I was still young and tender, my father introduced me to the greatest and most bitter of all lessons in life.

One fall day, as I returned home from an afternoon exploring a nearby forest, the door slammed behind me as I rushed into the house, eager to share my day's discoveries. My father was sitting on the three linoleum-covered steps that led to our kitchen. I sat at his side, excited to share with him the treasures I'd just collected. His normally ruddy complexion and handsome face were ashen, his brow drenched in sweat. Then I noticed the strange way he clutched his chest. The image transfixes me now: I, holding to my chest symbols of life—mosses, vines, lichens—and my father grasping his in the pain of a massive heart attack.

"Mike, leave Dad alone, he isn't feeling well," my mother said as she dialed for an ambulance.

The world of people I loved was now under attack by an enemy I couldn't comprehend.

A few months later, my mother and I walked hand-in-hand through the corridors of the Memorial Hospital in South Bend, Indiana, both of us carrying shopping bags of colorfully wrapped Christmas presents. When we reached my father's room, he put on a brave face and handed me an incredible gift. It was a biological oscilloscope, an electronic device that displayed the faint electrical charges in an organism as "blips" in a moving trace on a small green "television" screen. I knew what it had cost; I had coveted it for months. The price was some $350, a real sacrifice for a truck mechanic.

Late that night, I unpacked the wondrous instrument up in my laboratory alone. I smeared my wrists with conduction cream, unwrapped the electrodes, plugged myself in, and turned on the scope. There was a green dot, at first bouncing around randomly, then taking on a smooth and steady course across the screen. There in the slowly moving green fluorescent trace was my heartbeat. I had captured the "essence" of my own life on the oscilloscope. *How wonderful is this miracle we call life,* I thought to myself. This machine that had been made by people had inputted this life force, translated the energy into a light on a grid where one could measure the force of the energy. Could the true essence of life be captured that way? How desperately I wanted to understand that miracle, and give my father back an even greater gift.

My father recovered. In some ways, though, I'm not sure I ever did.

When my grandfather died, it fell to my father to arrange the funeral. My father had never attended church, but out of a deep sense of duty, he sought out a minister of the United Methodist Church. Either the loss of his father or the personality of the minister impacted him, because in the weeks that followed, he began to attend church regularly. I would often sit obediently at his side.

One Easter I turned over the church service program and read a quote by the founder of Protestantism, Martin Luther: "The Lord's promise of resurrection is written not in books alone, but in every leaf in springtime." *How true,* I thought. Life has an immortal force of renewal that we could actually see and feel. Out of the dead and frozen earth every year, the warmth of the sun brings forth new and green life. Luther suggested that this magnificent display of natural resurrection was a symbol, a promise, that God would in the same way bring us all forth again from the dust of death.

The Bible absorbed me as a sponge absorbs a drop of water. What a glorious solution to the problem of mortality. This, I realized, was the greatest force people feel drawing them to religion. It was a magnificent story of love conquering death. This is why the church was filled to brimming with a sea of gray hair. The audience was largely that sector of our population that was facing the inevitable blackness of the grave. As William James stated: "Religion in fact, for the great majority of our own race, *means* immortality, and nothing else. God is the producer of immortality."

My life turned about-face. With the same energy and single-mindedness that had characterized my earlier explorations in science, beginning in my last year of high school, I entered a period of passionate study of the Judeo-Christian tradition and its antecedents that would last a full ten years. I asked Christ to come into my life to be not only a savior but to be Lord as well—

that is to say, I offered him lordship, or day-to-day control over my life. The only rival for my love of God was my passion for science. Recognizing that science was my "first love," and the need to dethrone that potential idol, in a melodramatic yet nonetheless fervent gesture I burned my science books and demolished my old lab. On the empty bookshelves once holding tomes on vector calculus and quantum mechanics, I now stacked grammars of ancient Hebrew and Greek so as to study the Bible in its original languages, a life of Martin Luther, and works on the defense of the Christian faith.

But in a subtle way, I noticed even then that I differed from most of my Christian brethren. Many a time in witnessing to the unconverted on the street corner, or in a jail cell, I found myself saying, "Christianity is about truth. If you could show me evidence the Bible isn't true, I would cease to be a Christian." I meant it. I was always first and foremost on a quest for truth. And it was a promise easily voiced, as I had no doubt of the objective validity of the biblical message.

Striving to bring forth an inexpungable case for Christianity, a field called "Christian apologetics," I turned to a detailed study of the fossil history of life on earth. In my mind, the green essence of life, the force of the immortal renewal of life in spring, was identical to the person of Jesus of Nazareth. In other words, the answer to the central mystery of life has already been handed to us in the Bible. Like a problem in geometry where you are given the answer to a theorem and asked to fill in the blanks, step by step, to actually prove it, I just needed to find a way of proving that this was the case, or at least providing a rational basis for that belief.

I began with the argument from design. This line of reasoning holds that if we were to travel to another planet and were to find a watch, then there must have been a watchmaker. That is, the existence of an object of such complexity that it could not have occurred simply by chance suggests that it had a creator. I reasoned that it should be simple enough to distinguish between millions of years of gradual evolutionary change as predicted by Charles Darwin and a sudden creation of life a few thousand years ago as taught by much of modern conservative Christianity. These were, after all, very different explanations of the fossil record. Certainly, a careful study of the paleontological data should be able to distinguish between such disparate histories. And if the evidence really did point to a special creation, as numerous devoted Christians and men of science were saying, then any reasonable person would have to agree that we owe our existence to a deity, and that segued nicely to the answers offered by the Bible.

After finishing my bachelor's degree at Rensselaer Polytechnic Institute, I moved back to my hometown to work with my father and to study part-time toward a graduate degree in biology. I enrolled at Andrews University, a Seventh-Day Adventist institution in nearby Berrien Springs, Michigan, to earn a master's degree. Andrews had amassed a distinguished group of scientists who were devoted Christians and who also were deeply committed to finding the evidence for a creator in nature. Among them was Richard Ritland, who had studied theology but also earned his doctorate in comparative anatomy, studying under the venerated paleontologist Alfred Romer at Harvard. There was Ed Hare, who had earned a doctorate in geochronology at the California Institute of Technology, pioneering a new way of determining the age of fossils using amino acids, and Ed Lugenbeal, who had theology training but also was pursuing his doctorate in human physical anthropology at the University of Wisconsin at Madison. These and many other distinguished scientists taught out of George McCready Price Hall, a building named after the Seventh-Day Adventist who had laid the foundation of what we call modern-day creationism. The Adventists had long followed the dictum of their founder, Ellen G. White, who had taught that "true science is but an interpretation of the handwriting of God in the material world."

For the next five years I sifted through the data, looking to carefully document the fingerprints of the creator in the world of nature. It could easily have been a lifetime of hard work. But these generous scholars accelerated my education with many hours of personal interaction and field trips, taking the time to actually go look at the fossil record. To their credit, while earnest in their own desire to edify the Christian faithful, they were equally earnest to follow the truth regardless of its implications. Among all the scientists I would ever collaborate with, they were some of the finest. Ritland, in particular, was a treasure trove of knowledge of such miscellaneous fossil data as the London Shale and the fossil record in Yellowstone Park, and through it all, he always showed his steadfast love of truth over dogma.

I found my progress in understanding the fossil record rewarding, but I was not content with my success in defending biblical creationism. During those years I witnessed one cherished tenet of creationism after another shattered under the geologist's hammer. Each day I stood tall as I walked into Price Hall determined to defend the Bible; each day I left the building one inch deeper in the quicksand of evolution. I was progressively moving in the direction of the dreaded conclusion that the scientific establishment had it right. There was no vast conspiracy out there to undermine Christianity with a flood of misinformation. The scientists were just ordinary people who worked daily with fossils; they had accumulated a mountain of honest data in support of evolution. All I had to do to see this was take the time to break open the books and to read their scientific papers. To my chagrin, virtually all the evidence argued against the creationist position.

Toward the end of my tenure at Andrews, I had given up on the bulk of the theory of special creation. But I still clung to some hope that perhaps in the case of human origins there was evidence of our special status over and against nature. At that time I attended a seminar held by Ed Lugenbeal. After seminary, Ed had been sent by the Adventist Church to the University of Madison at Wisconsin to study the human fossil record and earn a doctorate in the field. Like me, his passionate goal was to find evidence of the creator. He had now finished his doctoral work, and was presenting the product of that research distilled into a one-hour seminar in George McCready Price Hall. I brought along my notebook, eager to hear all of the evidence against the evolutionary descent of man.

Much against my expectations, it was clear from the start that Ed was taking us in a different direction. He walked us through the sedimentology of the great Rift Valley in Africa, providing lucid and persuasive evidence that the human fossils were buried in sediments millions of years old. His case included very careful analysis of magnetic reversals, ancient pollen deposition, and the silt that accumulated in the depths of the ocean. In addition, he provided concrete examples of a progression of human fossils. In other words, in the rocks that were six million years old, there were primates but no hominids, in the fossils of six hundred thousand years ago, there were humanlike members of the genus *Homo* but not modern man, and so on. Ed's research had led to the logical conclusion that man had evolved over many millions of years. You could have heard a pin drop. As we filed out of the auditorium, we were all wearing long faces.

A few days later, I sought Ed out in the privacy of his home and asked him to retrace his logic in dating the human fossils. He carefully took me back through the data, pointing to where I could document this or that fact. His logic was unassailable. I suspect that only a devoted creationist could build such an air-tight case for human evolution. No one else would be so motivated. But if his goal had been to prove the establishment wrong and the Bible true, his despair now at seeing the data go in the opposite direction was palpable. He confided in me the depth of the scandal his research had spawned. George McCready Price was rolling in his grave. This is not exactly what Andrews had in mind when it sponsored his doctoral work. To make matters worse, he was now being ostracized by his own church, as though his moral integrity were in question.

My conversations with Ed were a turning point for me. I had been holding back on the belief that human beings had to be part of evolution. It had seemed to me that perhaps all other life had evolved but humans could have been separately created. Now, with Ed's thoughtful analysis, I had the distinct impression that any further studies in human evolution would be a waste of my time. I would come to the same conclusion he had come to. I knew I was now a believer in the general theory of evolution from top to bottom. I was finally convinced.

The day I admitted this to myself, I knew my life had changed. I had lost my moorings. My pen cannot find the words to communicate my feeling of betrayal. I had trusted those in the Christian community who were also scientists. I had faith that they were being straightforward and honest with the data. Now I knew that I had squandered years of my life chasing after an illusion. More important, I knew that there were implications to a belief in evolution, powerful implications, and now I had to face them head-on.

And then, at that worst possible moment, tragedy struck home. My father, in the hospital for a relatively innocuous procedure, suffered another sudden heart attack. My mother and I rushed to the hospital.

I noticed, out of the corner of my eye, standing at the door, a tired, white-coated doctor.

"I'm sorry, we did all we could do. . . ."

The rest was a blur.

One cold and rainy day that spring, I stopped at a drive-through for lunch on my way to class. I ate in my car, looking out at the rolling hills of our Silverbrook Cemetery. Somewhere in the far distance were the graves of my father and my grandfather. As I sat there, I thought of life as a paleontologist would—that is to say, from a perspective of the timeless millennia, looking down on the minuscule sliver of time we humans call a lifetime. I thought of the cycles of life and death, the mortality of individual beings, and the immortal thread of life that connects those generations.

There are two words in Greek that mean "life." *Bios* means "mortal life," and *zoë* means "immortal life." It is as if the individual fossil strata, the histories of the lives and ultimate deaths of individuals, are the *bios*. But *zoë*, as I knew, represents the immortal renewal of life. It was the force of life that gives rise to the origin of species. It is a substratum that lives on through generation after generation, transcending the millennia. It was the kind of immortality we have through our children. I am carrying within my beating heart the DNA of my father, and his father before him. It was the force represented by Easter, the bursting forth of new life, the ancient symbols of the lily, the crocus, the egg, and the fertile rabbit left over from the ancient mystery religions. It was the kind of life Jesus referred to when he said "I am the Life." There was an immortal basis of life, an immortal force

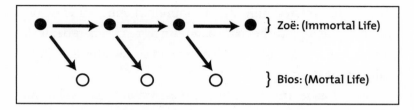

Figure 1. The Greek Conception of Mortal and Immortal Life. In ancient Greek thought, the immortal renewal of life was termed *zoë* and the mortal life of the individual was called *bios*.

connecting the lives of my grandfather, my father, and myself, a credible and visible evidence that in some mysterious way, life needn't always lead to death.

As Sir James Frazer taught in *The Golden Bough*, many cultures have valued their own version of the myth of immortality transcending the cycle of life and death. One destined to have the greatest impact on the ancient pre-Christian world was the myth of Osiris. This heroic god-man of the ancient Egyptians surely was not the first legend to illustrate the point, but it remains one of the most moving and, through the art and mythology of that great early culture, one of the most documented. According to legend, Osiris was a good king and taught his people the art of agriculture, in particular the cultivation of grain. Nevertheless, a jealous rival named Set plotted his murder. With the help of accomplices, Set tricked Osiris into lying in a coffin, whereupon he nailed it shut and plunged it into the Nile. Isis, Osiris's queen, recovered his body, but Set found it again and cut it into fourteen pieces, which he scattered throughout Egypt while Osiris's soul descended into the abode of the dead. Out of love for her husband, Isis diligently searched for the pieces of her beloved and, with the help of the gods, bandaged them together with strips of linen. Then she made love to Osiris. Through her love she resurrected him into an immortal spiritual existence. And because he was the first such resurrected human being, he was made the judge of those who would come after him in the realm of the dead.

So Osiris became a symbol of the immortal renewal of life. His face was typically painted green to symbolize immortal life, or black to symbolize the silt of the Nile from which such life arose. He was the symbol of life emerging from death over and over again in an eternal fugue. The ancient writer Plutarch said:

> Osiris is being buried at the time when the grain is
> sown and covered in the earth and then he comes to life
> and reappears when plants begin to sprout.

This same idea was reflected in what are called "Osiris beds." These were wooden frames in the profile of Osiris containing earth and seed placed in the tomb just as the seeds were in the process of germinating. They symbolized the hope that the deceased would also rise from the earth in a resurrected body possessing immortal life. As written in the Egyptian handbook to the afterlife, *The Book of the Dead*:

> Homage to thee, O my divine father Osiris, thou hast thy
> being with thy members. Thou didst not decay, thou didst
> not become worms, thou didst not diminish, thou didst not
> become corruption, thou didst not putrify, and thou didst
> not turn into worms. I am the god Khepera, and my mem-
> bers shall have an everlasting existence. I shall not decay, I
> shall not rot, I shall not putrify, I shall not turn into worms,
> and I shall not see corruption before the eye of the god Shu.
> I shall have my being; I shall have my being; I shall live, I
> shall live; I shall germinate, I shall germinate . . .

In a world united by the Greek and later Roman Empires, it could hardly have been accidental that Osiris and Isis had their counterparts elsewhere in the ancient world, at the eastern end of the Mediterranean Sea. Other peoples personified the concept of the immortal renewal of life under names such as Tammuz, Dionysus, Demeter, Persephone, Mithras, and of course, in the first Christian century, Jesus.

Many, I think mistakenly, call these fertility gods, as though their function in society were simply to provide a good crop or a profitable herd of cattle. As immortality gods, they certainly were the gods one would appeal to for a good harvest. But, in the moment that one faced death, it was the immortality gods, patterned after Osiris, face painted green to symbolize the essence of immortal life, that one looked for in the hope that, like the immortal cyclings of the sun and moon, like the reemergence of life in spring, so our loved ones would survive death to rise again, as surely as the sun on a new day. Like Osiris, they said, "I shall germinate!"

And I suspect the ancient Egyptians saw in the setting sun a metaphor for the end of life, and in the rising sun of the following morning the clear evidence that life will be born again after the darkness of night and death. The sun was a symbol of this force of immortal renewal, not merely an idol to be worshipped.

And as I sat there across the street from our hometown cemetery, it came to me with a powerful, almost irresistible force that the day would come when all the people I loved in this world, most of whom gave little thought to the cold reality of human mortality, would have their own names etched on tombstones on these very same hills. The day would dawn in which they would all be dead.

The sun would rise that day. Just as it had risen this morning.

I believe I saw clearly in that moment what few of us dare see. It has been said that no one can look directly at the sun or at death. But in that moment, I suddenly saw the human condition as I had never seen it before.

From deep within my soul, I erupted in an explosion of anger.

"This won't happen!" I shouted out loud at the thought of death.

This was the most profound experience of my life. I realized that it was simply not in my nature to accept death or be defeated by it. The call wasn't even a close one. I could never again resign myself to laying my loved ones down in the grave. It was crystal clear to me what I had to do. I had to defeat death. Was that even possible? In that fraction of a second when one sees a friend drowning and decides to try to save him, one doesn't really know if it is possible to save him or not.

This was the mission I'd sensed the day Albert Schweitzer died. It was to extract the "green essence" of life, the secret of the immortal renewal of life, to hold it in my hand and to give it to my fellow human being. This was the powerful synthesis I felt, the green face of Osiris, recapitulated in the great religions of the world, in my mind, equivalent to the biological basis of the immortal renewal of life. This was inarguably the greatest and highest calling of mankind, to find and control the biological basis of the immortality of life, and to alleviate the suffering of our fellow human beings. Within the meaning of the immortality of species, I had set in place a meaning of my own individual life.

I had a whole lifetime ahead of me. I was only twenty-six years old. I would dust off my scientific talents, resurrect my laboratory, and prepare myself for a long and hard-fought battle. I also knew enough about the scientific establishment to know that this particular path would make me a maverick. One of the

most eloquent and elegant science writers who ever picked up a
pen, Lewis Thomas, put the prevailing modern wisdom this way:

> *If we are not struck down, prematurely, by one or another*
> *of today's diseases, we live a certain length of time and*
> *then we die, and I doubt that medicine will ever gain a*
> *capacity to do anything much to modify this. I can see no*
> *reason for trying, and no hope of success anyway. At a*
> *certain age, it is in our nature to wear out, to come*
> *unhinged and to die, and that is that.*

> *No reason for trying.*
> *No hope of success anyway.*
> *It is in our nature to wear out.*

But that was wrong! And I knew with time I could demonstrate it.

Every instinct in me cried out that there was an immortal substratum of life, a *zoë*. It only remained for us to decipher it. Why
did the mortality of cells and therefore of human beings have to be
our lot? Where was it so written? Given all we had already done to
mold life on the planet to our needs, were we really forbidden to
apply our hands to the problem of human mortality?

This was a forceful experience that colored every waking
moment of my day. I was now a different person. Within two
years I had managed to sell off most of my father's businesses,
arranging in the process for my mother's well-being. Then, after
careful study, I had called the cell biologist Dr. Samuel Goldstein
at the University of Arkansas for Medical Sciences in Little Rock.
At the time, the Goldstein lab was the only group in the world
using advanced genetic techniques to study aging, and it was
there that I hoped to begin my quest. I remember him saying in

that phone conversation that, in order to understand human aging, he thought it best to study human cells (as opposed to fruit flies or worms). He was working on the "fibroblast model" of human aging. The fibroblast is a cell deep within the skin, easily accessible for research. His lab was committed to the study of the aging of these human cells in the laboratory dish and to finding the genetic clock that governed the process.

Goldstein, in turn, sounded impressed by what I had to say about my focus, and perhaps my conviction. He invited me down for an interview, and not long after that, I traveled to Little Rock to meet him and visit the university.

I found Sam to be a gentle and refined man. He was an M.D.-turned-researcher, precise in his choice of words, soft-spoken yet quick to respond with a spark of wit. He seemed pleased to find a rare soul interested in the biology of aging, and shortly thereafter I was admitted to the graduate program, with Sam himself as my adviser—and there was a dividend to my visit. By a stroke of good timing, I showed up in Little Rock during the famous Arkansas Creation Trial, a modern replaying of the 1925 Scopes "Monkey Trial" that had taken place in Dayton, Tennessee.

The trial of John Scopes, a high school biology teacher, had resulted from his teaching students Darwin's theory of evolution and thereby breaking the state's Butler Act forbidding "any theory that denies the story of the Divine Creation of man as taught in the Bible, and to teach instead that man has descended from a lower order of animals." The prosecution was led by the eloquent William Jennings Bryan, former secretary of state and presidential candidate.

In a reversal of roles, this new Arkansas trial (*McLean* vs. *Arkansas*, 1981) had been launched when the American Civil Liberties Union challenged a bill signed into law by former Arkansas governor Frank White that required equal time to creation models whenever evolution was taught in a public classroom. As a result, Governor White was portrayed in cartoons as a bumpkin with a banana, and there was a media circus that gave all of us in attendance that day an eerie sense that we were reliving the earlier trial.

In the Scopes case, the only living geologist that William Jennings Bryan could name that supported creationism was George McCready Price, the Seventh-Day Adventist who had inspired the Adventists' foray into twentieth-century creationism, the one for whom the Adventists had named the building at the school I attended in Michigan. Unfortunately, Price had had no formal training in geology, and it showed.

Now, in 1981, I managed to slip into the courtroom in time to see an old acquaintance of mine, Duane Gish, a Ph.D. biochemist, squirming in the witness chair as he sought to defend young-earth creationism. Painfully aware of the errors of his reasoning, I nevertheless felt sorry for him. His motivations were quite sincere and honorable—just dead wrong.

I heard a grunt and a snicker from the man sitting next to me, and, assuming they were expressions of dismissal, I said quietly, "You have to sympathize with Gish just the same. He's a biochemist by training, and a good one. Like George M. Price, he made the mistake of trusting his colleagues on their interpretation of the geological data."

But the man was implacable in his judgment.

"He dug his own grave," he pronounced gruffly. "He can lie in it."

"Yeah, but I spent ten years of my life digging the same grave, and I can sympathize with him."

"Well," he responded sardonically, "now you have a whole lifetime ahead of you to claw your way out!"

Only with the next recess and the onslaught of lights and television cameras approaching this man did I realize who he was. He was Stephen Jay Gould, the author and outspoken evolutionist. Of course, Gould was right, and I knew it. But still, in some ways my heart was more with Gish. Gish, though flat wrong about his "evidences for creation," was in some sense a kindred spirit. He and many of the creationists and evangelical Christians I knew were people of great compassion. Their motivations to lead others to the gospel of Christ, and their willingness to stand in opposition to opinions with which they differed, were in fact quite noble. It was only our focus that differed. While they acquiesced to the realities of aging and death in the hope of an eternal life in the world of spirit, they, like me, saw death as intrinsically evil. I, on the other hand, was taking a more pragmatic approach to doing good in *this* world. My goal was much more earthy and immediate. It was to wrest the secret of *zoë* using the power of the scientific method, to find the clock of aging in the ribbon of life we call DNA.

Chapter Three

Aging
Under Glass

Organic bodies are perishable; while life maintains the appearance of immortality in the constant succession of similar individuals, the individuals themselves pass away.
—JOHANNES MÜLLER

I sat down and peered into the microscope.

"Those are young cells," Sam explained. "They're like young people, narrow around the waist, busily occupying themselves with their favorite activity: reproduction."

It was August 1982, and my first day as a doctoral student in biochemistry at the University of Arkansas for Medical Sciences.

Sam Goldstein and I were side by side in the "Holy of Holies," the tissue culture suite. I considered it a "holy" site because in my imagination this room, circled by incubators, flashing lights, and gauges showing temperature and gas concentrations, held the answer to the mystery of life itself. Here in these incubators were young and old cells, and cells from diseases that cause premature aging, like progeria, a tragic illness that causes children to wrinkle and die when they are only teenagers. This room symbolized the challenge I had set for myself: to decipher the mystery of human aging and to design a strategy to defuse this ticking time bomb.

Sam swung the doors of an incubator open; inside were dozens of glistening sterile plastic dishes neatly piled in stacks of four, each with a quarter of an inch of clear red culture media not unlike cherry Kool-Aid in appearance. The red color did not come from blood, though the medium did contain blood from fetal cows. The red color actually came from a chemical called phenol red, a dye added to show how acidic the medium was. If it was red, the acidity was perfect. If it was yellow, it was too acid. If it became purple, then it was not acidic enough. He deftly extracted another dish, centered it on the microscope stage, and manipulated the knobs.

"Now, here are senescent cells," he said. "Take a look. Can you see any differences?"

"They're fatter," I observed, peering through the microscope.

"That's right. And slower moving. More 'spread out,' too. Do you see what I mean? Polygonal in shape. And like us when we get older, less and less interested in reproduction."

With a chuckle, Sam then left me alone. Surrounded by the humming scientific equipment, I settled in on what was to be the subject of my life's work. Finally I was back in my element, the world of science. I turned the knobs, carefully focusing on the cells. They reminded me more than anything of circus tents viewed from the air. They were totally unlike their slender young counterparts. There in the middle of each cell was the small oval structure, the *nucleus*, that harbored the DNA.

"What makes you old?" I asked them out loud.

I sat hunched over the microscope for nearly an hour, staring at the cells, gazing down at them in the hope of absorbing some knowledge, some sense, some instinct of where to begin in trying to find my way through this labyrinthine mystery. *What makes you old?* My mind raced over the alternative theories I'd learned up to that point. Most of them were "wear-and-tear" theories. That is, they proposed that our cells aged the same way cars age, or any other man-made machines: from a progressive accumulation of damage with time and use.

This was indeed a very commonsense explanation. Many young gerontologists, when charting the course of their career, gravitate to this kind of perspective. After all, virtually everything else wears out with time. The Second Law of Thermodynamics says that, in time, any isolated material must, in a sense, progressively deteriorate. The car is an excellent example. Its tires progressively lose rubber from friction with the road, its u-joints lose metal transmitting the force of the engine to the wheels, its brake shoes need relining, and so on. So why should human beings and human cells be any different? Or so the wear-and-tear theories proposed.

As commonsense as this sounds at first, it didn't ring true to me. In fact, it contradicted everything I knew about our evolutionary history. According to the modern version of Darwinian theory, life initially arose on this earth as a molecule that had the unique ability to take basic building blocks and assemble them into copies of itself. And the copies, of course, could do the same. Provided enough building blocks were available, the molecule could do this without limit. That is to say, it was *not* programmed to age.

Over millions of years, that original molecule, perhaps a molecule like RNA or DNA, found it advantageous in the struggle for existence to surround itself with a membrane to sequester the building blocks it had collected and to keep them away from the competition. Therefore, these reproduction machines became more and more competitive by becoming more and more complex, until eventually an organism similar to a cell came into being.

Simple single-cell animals, such as the pond-water organism called *Tetrahymena*, abound in our contemporary environment. They, too, like the original assemblies of molecules, are immortal. Unlike you and me, they leave no dead ancestors behind. By this I do not mean that they *cannot* die, only that they are not *programmed* to do so. Hit an immortal cell with a hammer and, yes, it will die. Starve it for food and it will die. "Immortal" in the above

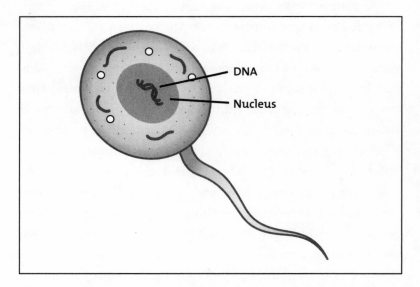

Figure 2. The Evolution of the Immortal Cell. In the continued competition for immortality, assemblages of molecules in the form of a cell evolved. These first cells, like the molecules from which they arose, were immortal.

sense really means that the cell is *potentially* immortal—that is, that there is nothing inlaid in its DNA that calls for aging—and that, barring accident, and provided it is surrounded with ample nutrition, it can go on living and reproducing forever.

The discovery of the cell had to await the invention of the microscope in the seventeenth century. When scientists turned this instrument on life itself, they saw a world of microscopic entities, cells that carried disease, cells like sperm that were the medium of sexual reproduction. This changed everything. A light shone down and the science of cell biology was born. The new dogma was summarized in Latin as *omnis cellula e cellula*—that is, "all cells arise from preexisting cells," a finding based on an earlier recognition of the continuity of life, that "all life arises from preexisting life."

It was only in the nineteenth century that scientists fully comprehended that humans are collections of trillions of cells mortared together like the bricks that make up a building. This cemented our understanding of how we develop in the womb, through the multiplication of a small ball of primitive cells and the unfolding, like the petals of a flower, of all of the different cells and tissues of the body. It provided a firm foundation on which to build an explanation of how living creatures reproduce themselves. And finally, the studies of the Estonian scientist Karl Ernst von Baer showed that humans reproduce by means of egg and sperm cells.

The staggering implications of cell theory captivated August Weismann, a brilliant German naturalist who taught zoology at the University of Freiburg. Understanding the implications of the new evolutionary biology and cell theory, he put the two pieces together and realized that life must have begun as immortal single-celled animals. It was Weismann who coined the term *immortality* for these cells. In his words:

Unicellular organisms . . . increase by means of fission. Each individual grows to a certain size and then divides into two parts, which are exactly alike in size and structure, so that it is impossible to decide whether one of them is younger or older than the other. Hence in a certain sense these organisms possess immortality . . . if protected from a violent death, they would live on indefinitely. . . . Each individual of any such unicellular species living on the earth today is far older than mankind, and is almost as old as life itself.

If we begin, therefore, with the simplest of organisms, these immortal single-celled animals, and assume that they enshrine the recipe for life (which we now know to be DNA), then it is simple to see that, by their passing a "photocopy" of the recipe on to their offspring cells as they divide, they maintain the so-called *germ line,* enabling it to proliferate forever with its precious DNA cargo intact down the generations and the eons.

Yes, and this is all very well for single-celled creatures. But what of multicelled ones like rabbits, trees, and you and me? How did they come about in the first place?

Weismann hypothesized that at some point in the evolution of life, certain unicellular individuals didn't separate immediately after division but lived together, at first as equivalent beings. As they grew, they divided again, making clusters of four cells, then eight cells, and so on. They were essentially the same immortal cells, but by sticking together through the challenges of life, they could reproduce better than cells that struck out on their own. Weismann gave an example of such an animal that still lives today. It is called *Pandorina morum.*

But then he pointed to another animal called *Volvox minor.*

In the case of this animal, some of the cells in the cluster changed ever so subtly. That is to say, the cells on the outside formed a kind of shell, dividing until the ball contained, say, two thousand cells, and then stopped dividing. At this point they died, releasing the original unaltered immortal cells so that they, in turn, proliferated and made a new *Volvox minor*.

Then Weismann made a revolutionary proposal.

He suggested that this shift in life strategy symbolized by *Volvox minor* likely represented the origins of aging and death. The cells that made up the shell were called "somatic" cells, *soma* being the Greek word for "body." The cells of the soma were "programmed" to self-destruct, having been used merely to feed, transport, and in general facilitate reproduction.

He called the immortal cells "germ line" cells from the French

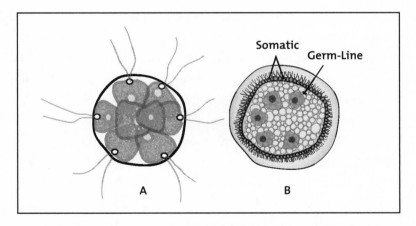

Figure 3. The Evolutionary Origin of Aging and Death. Weismann pointed out the subtle shift from simple multicellular animals where all the cells are immortal, such as *Pandorina morum* (a), to animals that were a mixture of mortal and immortal cells, such as *Volvox minor* (b). Weismann proposed that the germ line/soma dichotomy was the origin of the aging process.

word *germe,* which means "shoot" or "to germinate," as a seed sends forth a new shoot to form a new life.

As the great American naturalist and author Joseph Wood Krutch would later put it: "The amoeba and the paramecium are potentially immortal. . . . But for Volvox, death seems to be as inevitable as it is in a mouse or in a man. Volvox must die . . . because it had children and is no longer needed. When its time comes, it drops quietly to the bottom and joins its ancestors."

Now Weismann took the story one step further. Charles Darwin had a few years earlier published the seminal work *On the Origin of Species,* wherein he carefully laid out his theory of evolution. The word *evolution* literally means "an unrolling." Weismann proposed that the original immortal single-celled organisms have continued rolling on over millions of years, spinning off a soma every generation, the reproductive lineage itself being carried along on its immortal ride. Over eons of time, as recorded in the rock layers, through the Paleozoic Era (from the words "paleo," meaning ancient, and "zoë," meaning life), through the Mesozoic (middle life), and Cenozoic (recent life), spanning some billion years in time, hundreds of billions of days and nights, they evolved so as to make somatic cells that functioned as muscle to move arms and legs, into nerves to control the muscles, into a central processor called the brain to coordinate the motion, into eyes to see where the organism was moving, and so on. But like those first somatic cells of organisms like *Volvox,* all of the new complex somatic tissues (including those of human beings) were mortal. Evolutionary selection found no advantage for them to be anything else.

When life is viewed from this perspective, many things suddenly make sense. This worldview explains why nature goes through all the trouble of making a body only to have it die. Like the stages of a rocket that are cast off when their fuel is spent, the

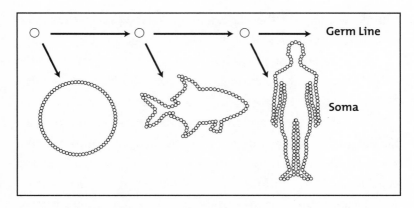

Figure 4. The Evolution of the Human Mortal Soma. The mortal soma of a sim-ple *Volvox*-like animal evolved over time into a more and more complex soma, though still maintaining the characteristic of mortality.

soma is merely a transport vehicle, a serf class of cells serving the reigning immortal germ line.

Weismann was moonstruck by this magical view of life. He could think of little else. How majestic it was that there were immortal cells that connected the generations, uniting us all like the wandering branches of a vine, of which each of us is an evanescent flower! _

It now became clear to Weismann that this was the simplest explanation of how hereditary information was passed on from generation to generation. The germ line was immortal and transported the genetic information in its nucleus to the next generation, while in every individual life cycle the genetic infor-mation in the soma simply died with the body.

Weismann's theory of the immortal germ line soon won out over the much more clumsy theory, advanced by Charles Darwin, that somehow the somatic cells send out bits of information that are assembled as a new germ cell in the testes and ovary.

Then Weismann asked what I think is one of the most profound questions in the history of biology: What is the actual clockwork that causes the human soma to age and die?

In September of 1881, Weismann delivered a lecture to his fellow scientists at the Association of German Naturalists titled *"Über die Dauer des Lebens,"* or "The Duration of Life." He made the astonishing prediction that, while the germ-line cells of multicellular animals (including human beings) were immortal, the somatic cells were intrinsically mortal—that is, they had the capacity to divide for only a finite number of divisions:

"Death takes place because worn-out tissue cannot forever renew itself, and because a capacity for increase by means of cell division is not everlasting but finite."

This simple proposition—that the capacity of cells to divide is not infinite but finite—would become a milestone in the history of gerontological research. It is all the more amazing in that, when Weismann offered it, neither DNA nor cultured cells had been studied in the laboratory to the point where his theory could be adequately tested. As he himself put it:

"I am unable to indicate the molecular and chemical properties of the cell upon which the duration of its power of reproduction depends: to ask this is to demand an explanation of the nature of heredity—a problem the solution of which may still occupy many generations of scientists."

Weismann was wonderfully, seemingly miraculously, clairvoyant. It would take half a century and the discovery of the structure of DNA in 1953 for James Watson to catch a glimpse of the actual mechanisms of heredity and immortality.

Unfortunately, Weismann's theories held sway for only three decades before they were seemingly discredited. In 1911, an illustrious surgeon and medical researcher at the Rockefeller Institute in New York, Alexis Carrel, and his partner, Montrose Burrows, pioneered a new technique called *tissue culture*. Carrel was the first scientist in the United States to be awarded the Nobel Prize, which he won for his work in perfecting methods to sew together two blood vessels. Now, having read reports of some success in coaxing cells into living in the artificial environment of the laboratory dish, he began to grow cells in a "liquid cuisine" (now we would call it a "culture medium") that contained all the nutrients the cells needed to grow and replicate outside of the body.

In their soon-to-be-famous experiment, Carrel and Burrows took cells from an embryonic chicken and placed them in glass dishes with culture medium. In this environment, the cells could live and indeed divide and increase in number over time. Because they were propagating in glass dishes, it was said that they were grown *in vitro* (from the Latin meaning "in glass") as opposed to *in vivo* (Latin for "in life"—that is, in the body). Carrel demonstrated to everyone's satisfaction that the cells thus cultured could proliferate for days, and indeed months, outside of the body.

So the stage was set for a test of Weismann's theory. The question Carrel asked was: Did the chicken cells age in the dish, or were they immortal when freed from the constraints of the animal?

After a period of time, Carrel cut some of the cells away and put them into a new dish so they would have room to continue dividing, a process we now call "passaging the cells." Carrel's culture thrived; it grew and grew. In 1912, he published his first results in a paper titled "On the Permanent Life of Tissues Outside of the Organism." Carrel reported that the cells showed no signs of aging, even after months of continuous passaging. He concluded that cells isolated from the chicken and cultured in vitro did not in fact age. In other words, Weismann's prediction was wrong.

He continued to report on the cells year after year, and the cultures proliferated for over thirty years before the experiments were terminated in 1946, two years after Carrel's death. They supposedly lived over three times the reported life span of a chicken—in other words, the rough equivalent of two hundred human years!

Of course, each time the culture was passaged some of the cells were thrown away, but, as Leonard Hayflick pointed out years later, there was great speculation of how many cells there would have been after thirty-four years of continued growth:

> Ebeling remarked that, if they had been saved over thirty-four years, the mass of chicken cells that would have accumulated would have been greater than that of the sun! In 1921 a New York newspaper called the World wrote that they would have formed a "rooster . . . big enough today to cross the Atlantic in a stride . . . so monstrous that when perched on this mundane sphere, the world, it would look like a weathercock."

Needless to say, Carrel's results delivered the coup de grâce to Weismann's patiently worked hypothesis. Carrel's paper was published in 1912, just two years before Weismann's death. Because of his status as a Nobel laureate, most of the scientific community quickly accepted Carrel's results. It soon became dogma that however we aged, it was not due to an intrinsic property of our cells. Our somatic cells, if provided proper nutrition, it now appeared, were immortal.

August Weismann died with much of his life's work discredited. His theory of the mortal somatic cells was buried in the dusty annals of forgotten scientific literature.

———————

But the story was far from over.

More than three decades later, on February 1, 1951, a thirty-year-old black woman named Henrietta Lacks drove to Johns Hopkins Hospital in Baltimore, Maryland, after some unusual intermenstrual bleeding. On inspecting her cervix, the rim of tissue at the back of the vagina that opens into the uterus, the gynecologist who examined her noted an unusual purple growth that he took to be a cancer. As is typical in such cases, he took a small sample of the mass to be studied under a microscope to see if it was indeed a malignant cancer or merely some kind of benign growth that would not need treatment. Before she left that day, Henrietta had her diagnosis. The growth was malignant. It was cancer. Eight days later, she repeated the visit; surgery was performed to remove the tumor, and radiation was applied to kill any remaining tumor cells. What she never knew was that a small piece of the tumor was secretly handed over to the tissue culture lab of George and Margaret Gey at the same institution.

The dream of Dr. Gey and his wife was to get cancer growing in the laboratory, to establish a "tumor in a test tube." They hoped that if cancer could be corralled and grown at will in tissue culture, scientists could begin a serious study of the ultimate causes of the disease and test anticancer therapies on cancer cells in the dish before trying them on people.

Dr. Gey's laboratory technician took the half-inch-sized piece of tumor tissue, minced it into tiny pieces with sterile razors, and placed it into glass bottles along with a "witches' brew" of clotted chicken blood, "juice" from an embryonic cow, and a dash of human blood. Within days, the observers in the Gey lab witnessed something no one had ever seen before. Whereas cells from many other tumors always shriveled and died within days of being put in the laboratory dish, Henrietta's cells began an explosive campaign of growth in this artificial environment. They sucked up the nutrients fed to them and, piling up within the dish, formed millions upon millions of new cells that were visible even to the naked eye as opaque disks on the inside surface of the culture vessels. Over the next few months, it became clear to Dr. Gey that he had an immortal human cell line made up of cancer cells, one that would grow over extended periods of time just like the chicken cells studied years earlier by Alexis Carrel.

Sadly, the treatment Henrietta received didn't help her. Within months, tumors appeared on almost every organ in her body, and she died on October 4, 1951. In a touching illustration of the contrast between mortality and immortality, on the very day when Henrietta died, while her body lay in the morgue and the pathologists inspected the host of solid tumors like glistening pearls attached to the organs of her abdomen, George Gey appeared on national television with a vial of Henrietta's cells, which he called HeLa cells to hide their true origin. He held them up to the camera, saying, "It is possible that, from fundamental studies such as these, we will be able to learn a way by which cancer can be completely wiped out."

But the notoriety of HeLa cells was only beginning. Their descendants were destined to travel to research facilities around the world. They would venture into space on the *Discoverer XVII* satellite, they would be placed near an early atom bomb blast to measure the effects of radiation on human cells, they would be used to grow and multiply a millionfold the dreaded polio virus, thereby speeding the development of a vaccine, and millions of the cells would even eventually be injected into the arms of prisoners in Ohio to see if cancer was contagious, like polio itself, and they continue to live to this day. These were truly immortal human somatic cells. And eventually, they would play a role in the quest to uncover the mechanisms of immortality itself.

The refreshing thing about science is that it is self-correcting. If there is a mistake in the scientific literature, someone, someday, will find it and relish in setting it right. After all, there are few things that give scientists as much pleasure as overturning dogma.

In the 1950s, improved means of growing human cells became critical for making vaccines and in studying cancer. At the Wistar Institute in Philadelphia, Leonard Hayflick and Paul Moorhead were trying to get clean cultures of human cells from fetal tissue for a project aimed at identifying tumor viruses.

Hayflick was having some luck getting cells growing in vitro. But to his consternation, he couldn't match what the Geys had managed to achieve years earlier—that is, to culture the cells continuously. In his hands, the cells would grow nicely for about fifty doublings and then stop dividing. In frustration, Hayflick walked into the Institute's lunchroom one day, complaining about the problem to his colleagues. One of them replied jokingly, "Well, Len, maybe your cells are just growing old!"

Could that be?

Hayflick began to suspect there was some truth in that off-the-cuff remark. Human cells might be mortal after all. And how scandalous that would be! Hayflick knew that some human cells were clearly immortal. HeLa cells were a striking example. But interestingly, the truly immortal cells were always abnormal. If one looked into the nucleus of immortal cancer cells under a

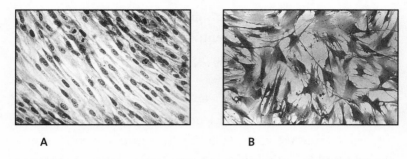

A   B

Figure 5. Young and Senescent Cells. (A) Young slender cells cultured by Dr. Hayflick that divide in response to growth signals. (B) Senescent cells that become enlarged and no longer reproduce, even when provided with nutrients that would normally promote cell division. (Photograph courtesy of Leonard Hayflick.)

microscope, one could inevitably see gross abnormalities such as broken or missing chromosomes. Perhaps, he reasoned, the cells of the body were normally mortal, and became immortal only after the genetic program was badly damaged, as in the case of cancer.

But old ideas die hard. How could Hayflick publish results that so completely contradicted the accepted scientific knowledge of his time? Other scientists suggested that perhaps he was just not culturing the cells right. In essence they were saying, "Why don't you come over to our lab and we'll show you the right way to do tissue culture!"

But if that was the problem, why was it that only those cultures that had achieved about fifty doublings were now failing to grow further? If there was a problem with the technique itself, then one would think that all the cells in the laboratory would be performing badly, those that had doubled only a few times as well as those that had doubled fifty times.

Hayflick decided instead on a series of definitive tests.

His first experiment determined whether the cells' senescence was due to the way they were being grown or was, in fact, an intrinsic property of the cells themselves. He tested this by mixing young and old cells together in the same dish. If senescence was caused by some problem in the way the cells were cultured, then the young and old cells should stop doubling at the same time, since they shared the same dish, were fed the same medium, and so on. If, however, senescence reflected an intrinsic "clock" of cell aging, then over time one would expect the older cells to stop dividing and the younger ones in the same dish to keep on going. Eventually, all the remaining cells would have been derived from the young cohorts.

To carry out the experiment, Hayflick needed a means to tell the original young cells from the old ones. There was no simple way of marking them, so he hit on the idea of using young female cells and old male cells, which could easily be distinguished by an innate marker, the sex chromosomes. Despite the loud protests of Hayflick's secretary regarding the lascivious mingling of an older man's cells with those of a young woman, the experiment was launched, a study later dubbed the "Dirty Old Man Experiment."

Hayflick discovered that the old man's cells stopped doubling as if on cue, while the young female cells continued to grow. To be doubly sure, he then reversed the conditions and did the analogous "Dirty Old Woman Experiment." The results were comparable. It appeared that cells did age after all, in response to internal cues. And nothing, not even the lewd cuddling with younger cells of the opposite sex, could delay the inevitable clock of aging.

Next he designed a test to determine whether his results were due to something peculiar to his lab. A basic tenet of science is that a result, in order to be believed and written in the textbooks, must be repeatable by others. So Hayflick sent his cells to the laboratories of other cell-culture experts and asked them all to grow the cells independently. After four months, the results were tabulated and sent back to Hayflick. The other researchers also reported having trouble growing his cells after a total of about fifty doublings.

This finally convinced Hayflick beyond any reasonable doubt—human cells *did* age. He decided to go public with his results. He typed up a manuscript and mailed it to the *Journal of Experimental Medicine*. Most scientific journals apply very stringent standards to articles submitted for publication. One of the ways they do this is to send them out to anonymous reviewers who are experts in that particular field, a procedure called "peer review." In Hayflick's case, one of the reviewers was Peyton Rous, a Nobel laureate and the journal's editor. After reading Hayflick's manuscript, Rous wrote back refusing to recommend the article for publication, saying: "The largest fact to have come out from tissue culture in the last fifty years is that cells inherently capable of multiplying will do so indefinitely if supplied with the right [conditions]."

Even great scientists like Rous can be blinded by convention. Hayflick's paper was eventually published in another journal and has since become one of the most frequently referenced papers in the history of medical research. Hayflick was right; Rous, Carrel, and the rest of the scientific community were wrong. And August Weismann had been right all along. As he had predicted some eighty years earlier, somatic cells had now been proved to be mortal. But unfortunately, enough time had passed that few people, even those in the gerontology community like Hayflick, knew of Weismann's prescient insights.

The phenomenon of cell aging is now frequently referred to as the "Hayflick phenomenon." It is usually represented as a curve rising and then leveling off. The ascent of the curve represents the increase in cell number over time in young proliferating cells. The slowing and leveling off of that ascent is the cellular aging process. Cancer cells, however, are abnormal in that they have an ability to divide without limit; that is, they are immortal. They are represented on a graph by a line that rockets into infinity—that is, the cells possess the capacity to increase in number over time forever.

Since Hayflick's original experiments, many kinds of somatic cell types have been grown in vitro. All those that will divide in the laboratory exhibit the Hayflick phenomenon. I say "those that will divide" because not all cells will proliferate in a dish. Many cells in the body—among them brain cells like neurons and heart muscle cells—are not capable of division in vitro and rarely, if ever, divide in the course of the aging process in vivo.

My new mentor, Sam Goldstein, had been pursuing the mysteries of cell aging through his investigation of an unfortunate genetic disease called progeria. Children with this mutated DNA are normal at birth, but then, at around the age of ten, begin to show symptoms of accelerated aging. By the age of twelve or so, they have gray hair, wrinkled skin, thinning bones; they usually die of a heart attack before they reach their twentieth birthday.

In 1969, Sam put skin cells from a patient with progeria in culture and showed that these cells, like the victim himself, aged at an accelerated rate. In contrast, he showed that the Galápagos tortoise (which can live hundreds of years) has cells that can divide far beyond the limits of human cells.

Subsequent research showed that the aging of cells was linked to cell division, not time itself. In other words, the mere

passage of time, or the normal machinery of life we call metabo-
lism, did not move the hands of the clock of aging. The clock
ticked only when the cells divided. And a young student in Len
Hayflick's lab named Woody Wright had shown that the swap-
ping of old nuclei into young cells, and vice versa, demonstrated
that the clock was to be found inside the nucleus.

This, then, was the state of our knowledge when I sat staring at
the senescent culture in our tissue-culture lab at the University
of Arkansas for Medical Sciences in 1982. The wear-and-tear
theories didn't ring true to me simply because they ignored the
fact that we were made from immortal cells that have managed
to repair damage as they rolled on throughout the era of the
dinosaurs and the Stone Age, for millions, indeed billions of
years, until they ended up in our parents' germ cells to make you
and me. At the core of life lies an immortal substratum, the germ
line. The germ-line cells that rolled on over the ages struck me as
being like the archetypal symbol of the scarab beetle, which
rolled a ball of its own eggs, a symbol of Osiris, the source of the
immortal renewal of life. Conceivably the soma, or body, might
age from wear and tear, but there is no *a priori* reason to assume
that was the case. All the body cells that facilitate the reproduction of the germ line carry the same genes that make the reproductive cells immortal. Perhaps they retain the precise repair
mechanisms of the germ-line cells. Instead of being a wear-and-
tear phenomenon, the aging process may reflect some far more
subtle and interesting modification of the DNA.

But if that was true, how could we decipher it?

I was still pondering these mysteries that day when Sam
Goldstein walked back into the lab.

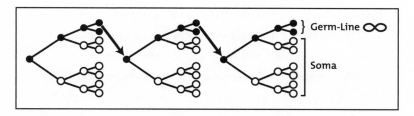

Figure 6. The Dichotomy of the Immortal Germ Line and the Mortal Soma. Hayflick's discovery of the aging of human somatic cells provided a laboratory model to study the biology of cell mortality and immortality.

"Your plate is purple," he remarked, pointing to the dish before me. It had been out of the incubator so long that the normal atmospheric gases had made the pH of the medium far too basic. "You may as well throw that one away."

I dropped the dish into an orange biohazard bag and returned to the workaday world of the graduate student. In truth, though, those cells had not been wasted. Contemplating them, and the scientific history behind them, set my mind on a productive course. Was there some kind of mysterious coding in the DNA? And if so, what was it? A kind of genetic switch? A mechanism that, when turned on—or turned off—in somatic cells millions of years before, had led to their aging and death, while the germ-line cells that perpetuated life, the *zoë*, remained immortal?

The idea was beautifully expressed in a book called *Jewish Insights on Death and Mourning* by Jack Riemer:

> *I am what I am because of the first amoeba that developed into a more complex form, impelled by the divine imperative to grow. . . . As I think of the long line stretching far into the past, I also cast my glance forward. The line into the future is just as unbroken. It moves through me into*

*generations yet unborn. And as I think of this, I am
comforted. For I am a point in that line, and the course
of existence travels through me. I have inherited from
all the past and I will bequeath to all the future. In the
movement of that line lies the secret of immortality, and
I am a part of it.*

It was that secret that I was determined to uncover.

When I finally left the research building that first day and
walked across the asphalt parking lot to my car, a flaring orange
sun was setting the clouds on fire. One more precious day
spent. That same sun was going to rise again, I was sure of that
much; but I was equally sure some of my fellow "mere mortals"
wouldn't.

There was a secret to *zoë*.

Someday, somehow, we would extract its essence in the test
tube.

Chapter Four

Ariadne's Thread

Ariadne's
Thread

It is those who know little, and not those who know much,
who so positively assert that this or that problem
will never be solved by science.
—CHARLES DARWIN

In a work called the *Theogeny*, the Greek author Hesiod described three daughters of the Night called the Fates. According to this legend, these feared goddesses delineated the length of the human life span. Simply put, they decided the day each of us is destined to die. The first goddess was Clotho, who spun the thread of each human life. The second was Lachesis, the

Figure 7. The Fates. Clotho (standing) spun the thread of life; her sister Lachesis drew it out and is pictured measuring it against a bough; and Atropos, shears in hand, cut the length to forever determine mankind's allotted life span. (Painting by Frederick Paul Thumann, 1834–1908.)

"apportioner of parts," who measured the length of the thread of life against her measuring stick. The third sister, named Atropos, the most dreaded of the sisters, cut the thread of life to its fated length.

It took over two thousand years, from the time of Hesiod to our own generation, for us to discover that the "Fates"—that is, the mechanism that measures and apportions the length of the threads of our lives, that makes certain of our cells mortal and others immortal—is in fact inside us all along, in our genetic code.

By the 1980s, a handful of would-be gerontologists had come to the conclusion that the answer to human aging was to be found in DNA. We converged like religious pilgrims on Sam Goldstein's lab. We all sensed that we were a few short years from discovering the clock.

By the time I joined the race in 1982, the tools to "cut and splice" DNA were already in place. Called "recombinant DNA technology," these new tools gave us the ability to do virtually anything with DNA that the human mind could imagine. We could now determine the sequence of the DNA of any species of plant or animal and even take that DNA and rearrange it any way we wanted. Put simply, we could now "tinker" with the blueprint of life. This new technology had caused an international furor over the power it conferred on mankind to manipulate nature, and bills were proposed that would have banned it outright. But the scientific community had made an eloquent case as to why research in this area should be allowed to proceed. By the early 1980s, the furor had largely subsided, and the technology was being applied in facilities around the world to address a host of medical research problems.

But how would we apply this powerful new technology to the mystery of aging? In 1977, scientists had discovered the means of chemically peeling away the letters of the DNA sequence one at a time, thereby allowing scientists to "sequence" the DNA. The "letters" in this case were the nucleotide building blocks of DNA. In each such nucleotide there are four bases: A (adenine),

C (cytosine), T (thymine), and G (guanine). If aging were the result of a change in the sequence of DNA, we could theoretically sequence the entire DNA of a young cell and then compare it letter by letter with the sequence of an old one. But the cost of that effort was clearly out of reach in 1982. (Eight years later, using far more advanced technologies, James Watson, the discoverer of the structure of DNA, launched the National Center for Human Genome Research to sequence the entire human genome. Still, it took hundreds of millions of dollars and thousands of people to achieve a single sequencing of all the three billion "letters" of the DNA, which was finally completed in 2001.)

So, in 1982, a complete search of the whole length of DNA to find changes associated with aging seemed daunting, to say the least. How else could we find the clock?

There are different categories of DNA we could study. There are forty-six strands of DNA in virtually all of the cells in the human body. Each of these strands, called *chromosomes*, needs to be copied when a cell divides; the copy is then transported to the new cell. The first category of DNA to consider is the gene itself. On each chromosome there are regions in the DNA that spell out what we call a "coding sequence." These are the genes. They are a code for making a particular thing, most often a particular protein, such as the collagen that makes the bulk of our skin, or the red protein complex called hemoglobin that is packed in our red blood cells, which serve to transport oxygen to where it is needed.

But there are other categories of DNA as well. Some serve a more structural role. For instance, there are regions on the chromosomes that serve as attachment points where cables can be tethered to help move the strands of DNA into their new locations in the daughter cells. These regions, often at the center of the chromosomes, are called *centromeres*, and their DNA usually contains no genes at all. Once called "junk DNA" (because they weren't genes and served no known purpose), they were now known to play a critical role in moving chromosomes. In 1980, Jack Szostack at the Harvard Medical School showed that the centromeres contain "repetitive sequences"—that is, they are constructed from stretches of DNA repeated over and over again; in the case of one sequence, it is repeated some 600,000 times per cell. This is most easily seen in the laboratory by taking DNA from cells and cutting it up with proteins that digest the DNA strands at particular sequences. Since centromeres have these stretches repeated hundreds of thousands of times, we get hundreds of thousands of pieces of DNA the exact same size. When we sort the DNA by size in a gel under an electric current, the large number of repeats migrate as a big "pile" of DNA easily seen as a glowing band on the gel. In the case of the centromeric repeats, they are easily observed after digesting the DNA with a protein called *Eco*RI (so named because it was found in the bacterium *E. coli*). They are sometimes referred to as the "*Eco*RI family of centromeric repeats."

Other repetitive sequences also exist in the DNA. One that occurs less frequently is called the "*Alu*" repeat sequence (also named after a protein that cut and revealed the repeats). Little is known about why the *Alu* repeat exists in our DNA; in 1982, there was a lot of speculation that perhaps *Alu* was a place in the DNA that initiated DNA replication.

Last, the linear ends of the chromosomes contain repetitive DNA as well. Early in the twentieth century, the scientist Hermann Müller focused on the strange properties of chromosome ends that he called "telomeres" (pronounced TEE-low-meers). The word "telomere" comes from the ancient word *telos,* meaning "end," and *mere,* meaning "parts." Müller published his findings in 1938. Since then the telomere has been shown to serve such functions as "capping" the ends of the chromosomes (think of the plastic ends of our shoelaces)—not so much to keep them from unraveling but to signal when there is damage to the DNA, such as a broken DNA strand. To use the shoestring analogy again, it would be as if we were using the plastic caps on our shoestrings to easily see if our shoestring was broken. By looking down and seeing two plastic caps, we would know immediately that there was no break.

So there were many kinds of DNA to study. Where would be the best place to start?

I was impressed by the fact that senescence, though inevitable for the majority of our cells, nevertheless was not *strictly* inevitable. Under some special conditions, such as when the cells were exposed to certain tumor viruses, the cells could gain the ability to proliferate without limit, just like the immortal germ line. The immortal HeLa cells I described earlier likely arose from just such a mechanism, a sexually transmitted tumor virus called *papillomavirus*.

Over many a beer we argued and debated the implications of this observation. One side in this debate (which included me) argued that senescence could not be due to the random loss of genes or irreparable DNA damage. Otherwise, how could a cell in the presence of a tumor virus so easily gain those genes back and become immortal?

It made more sense to think of senescence as a "genetic program." By this we merely meant that we believed there was a mechanism controlling the aging of the somatic cell. What was that mechanism? We speculated that perhaps it came from the loss of noncoding repetitive DNA. This would make the regions of DNA repeats a kind of clock, in the sense that a burning fuse is a "clock" or timing device. Because the progressive loss was in the DNA, it would provide a kind of "memory," allowing the cell to "remember" how old it was. We knew, for instance, that when cells at, say, fifteen doublings into their life span were frozen away for many years and then thawed again, they "remembered" how old they were. They picked up where they had left off and aged just as the original culture would have done if it had not been frozen. And a clock within repetitive DNA would help explain immortalization as well. As long as some of the repetitive DNA was still left in the old cell, immortalization from a tumor virus could result from the emergence of a mechanism to amplify the copies of the repeat sequence left, back to the original state.

But in 1982, no one could think of an easy way of studying telomeric repeats. A few years before I came to the lab, Sam and his partner, Robert Shmookler-Reis, had begun the process of comparing the amount of *Eco*RI repeats and *Alu* repeats in young and old cells. The data on *Eco*RI came in first. When Sam and Bob extracted DNA from young and old cells, cut it up with the *Eco*RI protein, and visualized the repeats on a gel, they saw that the band of repeats was dimmer in the DNA from old cells. The difference was subtle but reproducible—or so they thought. Excited that they might be closing in on the clock, they scrambled to write up their results and submitted a paper to *Cell*, a prestigious new journal dealing with the new DNA technologies.

After careful scrutiny and peer review from fellow anonymous scientists, the paper was published. Titled "Loss of Reiterated DNA Sequences During Serial Passage of Human Diploid Fibroblasts" and published in October 1980, it reported that every time a cell divided, about 1 percent of the *Eco*RI repeats were lost. Sam and Bob added that the probability that this result was merely chance was less than 1 in 30,000,000,000. I'll take those odds, I thought. I had a solid-enough foundation to start my own research work on the subject, and Sam agreed.

I decided to start by repeating Sam's and Bob's work with my own hands. One of the techniques they had described in their paper was called *saturation hybridization*—or "sat hybes," as we affectionately called them. It was the most precise means then known of comparing the amount of *Eco*RI repeats in young versus old DNA. This boring procedure cost me many an hour standing at a large machine that counted radioactivity on little circles of paper as I stared at the Little Rock skyline through the window. When the machine tallied the count, over and over again I saw no—absolutely no—difference between the young and old DNA.

What had I done wrong?

Young scientists learn to be critical of any and every result, but it took me a number of attempts, and an equal number of failures, for it to sink in that the *Cell* paper itself could have been wrong. Only by repeating the procedure and becoming more experienced in the techniques of working with DNA did I see where Sam and Bob might have made a mistake. When one extracts DNA from cells, one usually gets a lot of RNA as well. RNA and DNA differ only slightly in their structure, but RNA doesn't contain the *Eco*RI bands. In addition, while older cells contain the same amount of DNA as younger ones, they contain many times more RNA. It was possible, therefore, that the old DNA samples had more contaminating RNA than the young ones, and despite Bob and Sam's best efforts, the large amount of RNA in the old samples could have caused a miscalculation and an inadvertent underloading of DNA. If so, then perhaps there hadn't been a loss of *Eco*RI repeats after all.

But I was soon to learn that knowledge is not necessarily warmly received, even by objective scientists.

Shortly after finishing these experiments in the fall of 1983 and reporting them to Sam, I visited the laboratory of another scientist, who in turn introduced me to a young graduate student. She, too, had read Sam and Bob's *Cell* paper, and like me she'd embarked on a dissertation project to further explore this new finding. I was horrified. You see, the work of a biomedical graduate student is captive slave labor. The graduate student in medical research spends endless hours (including weekends) slogging through the tedious long-term experiments. Many times my colleagues would come into the lab at three in the morning to continue their own work and find me still at mine. The couches in the hallways were always full of students trying to catch a few hours of sleep before their gel had run its course. You fill in your notebook and clean up the radioactive waste, feeling the numbness caused by radiation burns on your fingertips. You put up with this thankless, backbreaking course because your Ph.D.—your whole future—is based on it. I knew what this young student was going through, and I knew she was wasting her time—and perhaps her career.

When I got back to Little Rock, I filled Sam in on the other student and her project:

"Sam, she hasn't the foggiest idea that the data may not be real."

"You didn't tell her, then?" Sam asked.

"No, of course not. Not before talking to you first," I answered.

"Did you tell her adviser?"

"No, I didn't tell anyone about it. I thought perhaps you should call them."

"Well, let's just leave well enough alone," Sam replied.

"But, Sam, she might be wasting months going down a path that leads nowhere."

"Well," he said with a shrug, "if you're right, then she'll figure it out for herself soon enough."

Troubled as I was by his seemingly cavalier attitude, Sam's intellectual prowess made such failings easy to forgive. In so many ways, he was a wonderful man. Frequently he invited us graduate students to his house, where we spent many enjoyable hours debating the evidence for this and that theory of cellular aging. Sam always relished the give-and-take of such speculation. But we kept coming back to the idea of structural changes in the DNA. It was the only reasonable explanation for the "clock" of cell aging. The numbers had to be somehow "written" in the length of the DNA molecule. But where?

Shortly after I'd arrived in the lab, Sam and Bob had submitted for publication an even more dramatic paper. Published on February 3, 1983, in the journal *Nature*, which had published Jim Watson and Francis Crick's paper on the structure of DNA thirty years earlier, Sam reported a striking instability in the *Alu* repeats in the aging process. The paper, titled "Extrachromosomal Circular Copies of an 'Inter-*Alu*' Unstable Sequence in Human DNA Are Amplified During In Vitro and In Vivo Aging," shook our little world of gerontology, if not the greater world of science. It reported that DNA in the middle of *Alu* repeats (hence "inter-*Alu*") was unstable in old cells and "popped out" of the chromosomes as small circles of DNA that were presumably then lost forever.

This was just the type of thing we had been looking for. Conceivably, as this DNA was progressively lost, the cells would eventually become dysfunctional and stop dividing. In my mind, the important question was why these inter-*Alu* circles were appearing in old cells. Could it be, I wondered, that the *Alu* repeats functioned as a place where the process of copying the DNA strands began? And, therefore, that in old cells there was a problem in getting DNA replication up and going? I imagined that replication tried to start, only to misfire, not unlike the way an old engine misfires as it tries and fails to start. The short strands of DNA made in these aborted attempts would then be shuffled off, or so I thought, as little circles.

I organized my theory and presented it one day over lunch to the lab. I cluttered the blackboard with data supporting my ideas and with proposals for a series of experiments to put these ideas to a definitive test. After the lunch, Sam led me into his office, closed the door, and with palpable excitement said, "I think your proposal is wild and wonderful."

By March 1984, I had accumulated a fair amount of data and the results confirmed my original findings; I thought I could now support the theory that replication misfiring occurred in old cells, and that Sam's *Nature* paper was just a small piece of a much larger picture. But one afternoon I decided that I had waited too long to run what is called a *control.*

The scientific enterprise is organized to discover truth and expose that which is false. That is why we send our papers to be anonymously reviewed by our peers, and that is why we do controls. A control is an experiment done to test the validity of the experiment of interest—for example, giving one group of patients a promising new medicine and the other a placebo. In my case, I asked a simple question: Are the circles of DNA I am looking at indeed human, rather than bacterial contamination?—and I used a bacterial probe as a control.

A *probe* is a piece of DNA made artificially radioactive. It is then mixed with a sample of DNA, and under the right conditions, it will stick to the DNA with the corresponding sequence. So when the DNA of interest is separated by size and attached to paper, the addition of the radioactive probe will make the DNA of a particular sequence radioactive. Since the target DNA is separated by size, the result is often a "band" or line on X-ray film that has been exposed to the paper. In this experiment, I knew the bands corresponding to the circles should not light up with a bacterial control probe, because they should not have had sequences in common. The bands corresponding to the circles should have lit up only with my replicon misfiring probe.

So I performed the control, then walked the X-ray film to the radiology department in our adjacent hospital to run it through the developing machine. I unpacked the film from its light-tight container in a darkroom, inserted it into an automated developer, and then walked outside and waited for the developed film to drop into a receiving tray. As I waited, I thought about my dissertation. I had already passed my candidacy exams for biochemistry. All I needed for my doctoral degree was to follow through on a number of experiments, write up my work for publication in a scientific journal, and write the more detailed doctoral dissertation.

As I stood waiting, the physicians around me stared at X-ray films of chests and ankles, and I realized how little I really knew about medicine. All of my studies were focused on the aging of cells. What did I know about the aging of people? Perhaps I should drop the idea of a postdoctoral fellowship and go to medical school.

The film dropped into the tray with a thump. I retrieved it and held it up to the ceiling lights.

"What the heck?" I exclaimed aloud.

The circles lit up with the bacterial probe only, and not with the human! Sometimes experiments don't work and you get crazy results that you never do understand. If you repeat the experiment and the quirky result never appears again, you just forget about it. I must have done something wrong, I assured myself. I went back to the lab that night and repeated the whole day's work, finally placing the DNA and X-ray film into an ultra-cold freezer to expose. After twenty-four hours I took the X-ray film out of the freezer, walked the frosted cassette back into the hospital, and waited again for my film to develop.

When the film was ready, I held my breath and again lifted it up to the fluorescent lights.

There it was again! The bands on the film were screaming out to me that the basis of what I hoped would be my doctoral research was merely bacterial contamination!

It had been a long week. I decided to go home and lick my wounds. When I got into bed, I tossed and turned, repeating in my mind every detail of the experiment. If the circles were merely bacterial contamination, why were they present only in the senescent samples?

I thought back to Sam and Bob's now-famous *Nature* paper. Had they run these same controls? I remembered that the circles reported in the *Nature* paper were 4.8 thousand base pairs in length—just the size one would predict if they were bacterial contamination!

Sam was out of town the next day, so when I got to the lab, I immediately knocked on Bob's door.

"Bob," I asked him, "the inter-*Alu* circles were 4.8 thousand base pairs, right?"

"Right," he said.

"Well, the bacterial probe itself is 4.8. Was there a control done to make sure the circles weren't just contamination?"

I put the question gently. I was scarcely prepared for Bob's reaction.

"Of course we did the controls!" he said. "The reviewers wanted them done, so we did them!"

"And they were negative?" I asked meekly.

"Of course they were negative!" His tone now showed that he was irritated.

At first I accepted Bob's word on this, but the coincidence in the size of the circles bothered me. Many times in my life I had been labeled a maverick. Perhaps I am, because I decided to do a somewhat rebellious experiment. The DNA samples used for the *Nature* paper were still being stored in our cold room, bound to a type of filter paper called nitrocellulose. I came into the lab late that night and dropped the famous filter into a steaming beaker of detergent that stripped off radioactivity. Then I placed this "scrubbed" filter into a plastic bag with a control probe and dipped it into a water bath to incubate for several hours. As I had before, I exposed the filter to X-ray film and put the film in the automated developer.

Thump.

Nervously I held the X-ray film up to the light.

Oh, my God!

The next morning I got in early. I watched for Bob to walk into his office and then knocked on his door.

"Bob, I've got some new results that I think are important," I said nervously. Bob followed me into the lab, where, perched against a light box, was the X-ray film.

"I probed the filter used in the *Nature* paper with pACYC184," I explained, "and the extrachromosomal circles all light up."

His response floored me.

"Oh, so I guess it was just artifact after all," he said in the most dismissive and matter-of-fact of tones. Then he left the room.

I couldn't understand. How could he be so blasé about this? I had just disproved one of the foundations of the lab's research. Why didn't he ask me about the details?

I waited till Sam returned, then asked for a meeting with him.

I must say that at first, after I'd carefully laid out the story, Sam was devastated. But then, recovering, he started in on a litany of questions: How did I know my results were real? What exactly had I done? What were the hybridization conditions? I'd taken his filter without asking him? How could I have done that?

And so on, as if somehow trying to prove that *I* was the one in the wrong.

In his eyes, I was clearly not "wild and wonderful" anymore. I was just wild and dangerous. But my findings, as the next days' and weeks' investigations proved, were unshakable, irrefutable.

I was hardly the first whistle-blower to be made to feel like the guilty party. Sure, I could understand his resistance to accepting the data for what they were, but how could he not appreciate my own predicament? By now, I'd been in the program over two years—time I'd spent, it seemed clear, chasing a ghost. If his findings on the *Alu* repeats didn't hold, how was I going to earn my Ph.D.? Wouldn't I have to start all over with a new research project?

On December 11, 1984, I went to see Sam. With all the conviction, and perhaps foolhardiness, of youth, I said, "Sam, you know I can't continue to work on my project in these circumstances. There are no circles—nothing to explain. I need—we need—to find something new for me to work on."

You could have cut the tension in the air with a knife. And then, to my mortification, Sam blew up at me.

"No," he said, waving me off. "You're going to continue to work on this problem. I am the adviser and you are the student. There is much more work to be done on this project."

"But, Sam, there's nothing to work on! It's not real. We—"

"We don't know that for sure," he interrupted.

"But *I* do. I *do* know it for sure," I said. "It's not *real*!"

At first there was dead silence. Then Sam said, "Well, either you're going to do what I tell you to do, or, if you're not willing to, then you can go find another graduate adviser."

It is hard to know the devastating impact of such a statement if you haven't gone through the trenches of graduate work—the thousands of hours of work. But as it seemed to me then, he left me no choice. On January 28, 1985, I wrote Sam a letter expressing my appreciation for the experience of being his student, left it on his office chair, collected my things, and was gone.

In hindsight, I know I should have given Sam more time. He had found himself caught unexpectedly in a very difficult situation—one that could have happened to any of us. And sadly, my leaving didn't make his problem any easier. It brought attention to the issue in the university administration. In the end, it led to Sam's writing a full retraction of the inter-*Alu* paper in *Nature*—about as humbling an event as can befall a serious scientist.

In the end, Sam kept his lab and his position. This was as it should have been, for he was as decent and compassionate a figure as I've encountered in my profession. As for me, I was obliged to begin my Ph.D. all over again, this time at Baylor Medical School in the lab of Jim Smith. Jim had trained with Leonard Hayflick, and along with his wife, Olivia Pereira-Smith, had a long-term interest in the biology of cellular aging. I formulated a straightforward study of the changes in gene expression during cellular aging that would give me my degree in four more years.

As I closed in on my Ph.D., I thought I understood the basics of how aging cells lead to aging tissues and aging people. At the same time, I knew little about the anatomy and physiology of the human body—and little, too, about the actual diseases associated with aging. So I decided to continue my education in medical school. The morning after running my last gel at Baylor, I enrolled at the University of Texas Southwestern Medical School at Dallas. These were the days when my mother would speak of me—with reason—as her son, the professional student.

The medical school I attended in Dallas in 1989 was a home for "gunners." A gunner was a student with his or her sights set on getting high grades and being the "top gun"—the student with the highest grade point average of the graduating class, which led quasi-automatically to the best offers for residency. My interests were quite different. I wanted knowledge—not only in cell biology and biochemistry, which I already knew quite well, but in microbiology, histology, pathology, physiology, endocrinology, neurobiology, pharmacology, and, of course, the dreaded gross anatomy.

I say "dreaded" not so much because of the apprehension of dissecting a fellow human being, but because of the horrendous task of memorizing the many hundreds of nerves, muscles, and miscellaneous components of the human body. The day the class in gross anatomy started, we assembled in the auditorium and listened to a lecture about our behavior in the class. "There will be one cadaver for every four students," the instructor said matter-of-factly. But then his demeanor changed.

"Your cadaver may not look like a human being," he said. "It is shaved of all body hair and preserved in formaldehyde. But it should be treated with the highest standards of respect. These were once people—people that passed you on the streets of Dallas. In many cases, I spoke to these people before they donated their bodies to science. They expect you to value their contribution to your career, and to learn as much as you can in order to one day alleviate the suffering of other human beings. Any student found not treating a cadaver with the highest standards of respect will be disciplined, perhaps even face expulsion."

We left the auditorium and walked single-file to the gross anatomy labs, surrounded on either side by the previous year's class, hooting and cheering us on. Minutes later, we descended into the basement of the student building and gathered around long rectangular stainless-steel boxes. On cue, we opened the doors on top of the boxes and peered inside. There, lying on a tray, under a layer of formaldehyde, was a naked human body. We turned a crank, and slowly, inch by inch, the greenish cadaver rose from the depths.

The first incision was the hardest. Laying a scalpel to a human chest felt like an act of violence. You had to override your natural feelings that this was evil. But as the weeks passed and our cadaver became a pile of tissue, the cutting became second nature.

Before I came to medical school in Dallas, I had heard of a professor there who had done his Ph.D. with Len Hayflick at Stanford. Within the first few weeks, I stopped by the office door of Woodring E. Wright, M.D., Ph.D.—or "Woody," as he was known. Woody struck me as a fascinating person. His office wall was covered with photographs of gargoyles, his blackboard crammed with plans for his experiments in muscle biology.

I explained to him my interest in aging, and we immediately hit it off. Whereas his lab was focused largely on muscle biology, he told me that his passion was cellular aging. He thought that the field of aging research was not yet mature; he had chosen to work on muscle biology until the right moment came, at which time he would shift his focus to gerontological research.

To my delight, he agreed to let me take over an unused room to set up my own little lab. I felt as if I had a new coal-room lab in the middle of medical school, and it gave me a wonderful distraction from the tedium of classwork. I could take time out during lunch to run an experiment or two. I was particularly interested in a type of "extreme senescence" Woody was studying called mortality 2, which occurs when you extend the life span of cells beyond the normal Hayflick limit without immortalizing them.

In time I reached the demanding third year of medical school. Third year is in many ways the most stressful time for a medical student. The first two years are for the most part class work. The third is intensive training on the wards with live patients, and the hours are medieval torture. My typical program involved a hectic dance of running from floor to floor for thirty-six straight hours, with only a few minutes to grab a bite to eat, then eight hours of sleep, a twelve-hour day, more sleep, then another thirty-six-hour stretch.

On call at 11:00 one night, I got a new patient, an elderly woman presenting with a sudden onset of confusion and now lying before me in a coma. My resident gave the preliminary diagnosis of hepatic coma and then abandoned her to take care of what he considered a more urgent case, a young girl with the flu. I wouldn't second-guess his judgment, but if medical school demonstrated one thing to me, it was the rampant bias against treating the old. I'd already had one such experience when I came across an elderly man in the hallway of the emergency room who'd been lying on a gurney for four hours. Though he wasn't my patient, I stopped to check on him. He was dead. Admitted for angina, he'd been set aside in favor of other patients and had quietly slipped through the cracks—slipped into the oblivion of death. The justification in that case was "Look, he was geriatric, he would have probably died anyway."

My female patient in the coma this night was a "geriatric," but her son didn't see her that way. He loved her dearly and pleaded with me to help her. I was merely a student, but I decided to give it a try. The library was open for physicians twenty-four hours a day, and so I went over to read up on the disorder. I found a recent publication arguing that a common contributing factor can be prolonged constipation. I trudged back to the wards, found the woman's son by her side, and asked him, "Is there any chance she is constipated?"

"Yeah, she was, and I went to the store to get her a laxative, but it didn't help," he said.

Well, there was one way to find out. I asked her son to leave, and I gloved up and gently probed her rectum. There it was, a baseball-sized hard mass of stool. Proud of my detective work and eager to help the poor woman, I stood there for hours poking and prodding to get it out and then—eureka! An absolute flood of stool, filling the bed—what a way to befriend the nursing staff! There alone with her at four in the morning, I got my reward. To my surprise, though she was still in a coma, a distinctive smile spread across her face. And, I confess, I smiled back, then promptly left the room to report a bed change.

To me, this was what medicine was all about. It was more about helping your fellow human beings when they are in need, especially the elderly, who have so much to offer society. It had less to do with the personal reward of "snap cures," easy fixes that make you look good as a doctor, cures that happen far more frequently in the young than the elderly. The woman was transferred into the intensive-care unit shortly thereafter and recovered fully.

The next morning after rounds, still flush with my success, I stopped by Woody Wright's lab.

"What do you think about the telomere paper in *Science*?" he asked.

I'd seen the paper. It was a collaboration between one of Sam Goldstein's old students, Calvin Harley, and Carol Greider, a young scientist known for her work on telomeres. They reported that there was a loss of telomeric repeats in aging cells, not unlike Sam Goldstein's work with centromeric *Eco*RI and inter-*Alu* repeat loss, the work I had previously refuted.

"It's just crap," I answered. "I'm really tired of these reports of loss of repetitive DNA in aging cells. These people just don't know the data. I know about Cal Harley, too. He worked with Sam Goldstein and published a paper with him on the *Eco*RI and inter-*Alu* story, and you know what I think of that whole fiasco." Even as I finished saying this, I knew I was overreacting. The Harley-Greider data were probably honest. If I had seen it during my years in Sam's lab, I would have probably chosen it as my own line of research instead of pursuing inter-*Alu*.

As I went back to the wards and in the days following, I ruminated on the telomere question. Back in Sam's lab, we used to debate the theory that the aging clock was in the repetitive DNA at the ends of the chromosomes, called the telomeres. In fact, the hypothesis had first been proposed by a young Russian scientist, Alexy M. Olovnikov, in the late 1960s. Then a student at the Gamelaya Institute of the Academy of Sciences of the USSR, Olovnikov heard a lecture at Moscow University by Alexander Y. Friedenstein, a cell biologist at the university, who held forth on Leonard Hayflick's discoveries that human somatic cells have a finite life span. As Olovnikov himself recalled it years later:

> *I was simply thunderstruck by the novelty and beauty of the Hayflick limit. I thought about this as I returned home from the University and walked along the quiet Moscow streets that were paved with gold-colored leaves on that early evening in late fall as I made my way to the subway station.*

What then occurred is one of those rare examples of insight that are as difficult to explain as the stroke of imaginative genius. As he waited at the subway station, Olovnikov focused on the Hayflick limit. "Hayflick's finding," he would write later, "was for me like Ariadne's thread was for Theseus, who followed it to escape from the labyrinth." Then he heard the roar of an oncoming train, and as the train came out of the tunnel, he realized that the tracks were reminiscent of the double-stranded DNA molecule. Every time a cell divides, the DNA double strand separates and a new complementary strand is manufactured for each half, thereby making two new DNA strands, one going into each new cell. The train itself reminded Olovnikov of a machine that rides along the DNA strands and manufactures the new complementary strands.

When the train stopped and the doors opened, Olovnikov noticed that no one got on at the very end. People got on *near* the end. Perhaps in the same way, the nucleotide building blocks (A,C,T,G) could not get onto the copying machine to replicate DNA at the very end of the telomeres. As a result, after every round of cell division and copying of the DNA strand, the end, or telo-

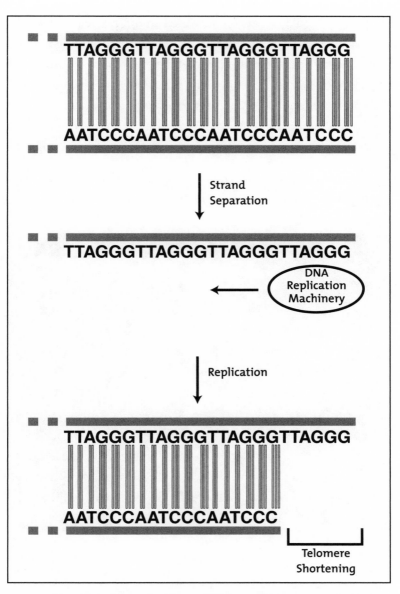

Figure 8. Olovnikov's Theory. The DNA replication machinery in mortal cells does not allow the building blocks of DNA to be incorporated at the very end of the newly replicated strands. As a result, telomeres shorten during cell division.

mere, shortened by a small but defined amount. The result would be that the process of cell division would continue without a problem for some period of time, but that the progressive shortening of the telomere over time would eventually cause the reproduced cells to experience severe DNA damage, leading to cellular aging.

In the case of the immortal germ-line cells, Olovnikov proposed that a special, yet-to-be-discovered type of DNA copying machinery existed that could replicate the very ends as well as the rest of the strands. In these cells, the DNA could be faithfully

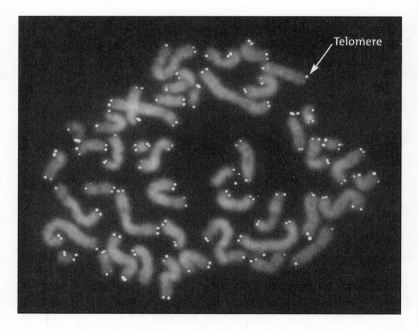

Figure 9. The Human Telomere. The blueprint of life is written in the molecule of DNA. It is present in individual strands called chromosomes (lightly illuminated). The ends, called telomeres, are visualized using a probe that specifically recognizes telomeric repeat DNA sequences and makes it brightly illuminated. (From young blood cells and courtesy of Peter Lansdorp, Terry Fox Laboratory, B.C. Cancer Research Center, UBC, Vancouver.)

replicated every time a cell divided, allowing the cells to live, potentially, forever.

Olovnikov published his telomere hypothesis in 1971 in a Russian journal, proposing for the first time in history a concrete theory of how Clotho spun the thread of life and how Atropos cut it to limit the life span of somatic cells. As significant as his hypothesis was, though, it lacked detail on the exact mechanisms at work. Less than a year after that first paper, James Watson, the codiscoverer of the structure of DNA, offered further insight. He observed that one end strand of DNA, called the "lagging strand" because of the way DNA replication occurs there, had no known means of being replicated at the very end. Since many more scientists read Watson's paper than Olovnikov's, Watson's was the one widely discussed, and the issue came to be known as Watson's "end replication problem."

Watson has denied knowing about Olovnikov's paper in 1972, saying, "I have come up with two good ideas in my life, and the end replication problem was one of them." Nonetheless, in a world where being the first to publish is of such paramount importance, Olovnikov should certainly share in the credit. At the time he wrote, though, perhaps because his paper was in Russian, still more because his ideas were so ahead of his time that there was no way of testing them in the laboratory, few scientists, even among those who were aware of what he'd written, took him seriously. The exception would have been a very small cadre of gerontologists, including myself.

The night after my conversation with Woody about the Harley-Greider paper in *Science*, I tossed and turned in my bed. My mind was racing, in turmoil. Could it be that I was overreacting in rejecting the telomere data? What were the actual facts of the matter? How could I rationally rule out Olovnikov's proposal without looking carefully at the current data?

Then I sat up in bed as if struck by a bolt of lightning. The telomere! Like putting a jigsaw puzzle together and remembering a bit of color or pattern that should go right *there*, I suddenly remembered a miscellaneous observation about the super-old cells Woody, Jerry, and I had been working with. Cells in mortality 2—the experiment where we tricked cells into advancing to extreme old age—showed the odd characteristic that almost every cell had two chromosomes fused together at the ends (or telomeres). If the telomere hypothesis were correct, then it would explain this strange result! Perhaps the shortening of telomeres could lead to cellular aging (mortality 1). And if we artificially forced cells into extreme old age (mortality 2), perhaps more than one telomere would be damaged. In that case, the cell would logically put these two damaged ends together, perhaps as part of a repair pathway, as though it were reassembling the two ends of a broken chromosome.

The more I thought it over, the more pieces fell into place. As much as I needed sleep, I needed answers more. I threw on my clothes and drove the short drive back to the medical school library. I pored over rows and rows of bound scientific journals, lugging heavy piles to the photocopying machine, copying every paper I could find on the subject of telomeres. Before the night was over, I had a good overview of the history of telomere research. It was clear that the lengths of telomeres differed from cell to cell, but no scientist who'd investigated the question was sold on Olovnikov's theory.

In fact, it wasn't until 1978, when Dr. Elizabeth Blackburn and Joseph Gall, then at Yale University, obtained the DNA

sequence of the telomere of the immortal single-celled pro-
tozoan called *Tetrahymena*, that the light began to dawn on
our modern understanding of the ends of the DNA strands.
Blackburn and Gall showed that the *Tetrahymena* telomere
sequence was TTGGGG repeated over and over again—that is, it
was TTGGGGTTGGGGTTGGGGTTGGGG, or as scientists say,
$(TTGGGG)_n$, where n is some unspecified number of times the
sequence is repeated.

This new information posed some interesting questions for
scientists. Why this sequence and not some other? Why were
there such sequences at the ends at all? Why was the sequence
repeated over and over, and how was the repeat generated in the
first place? How did telomeric repeats compare from one species
to another?

Dr. Blackburn focused on the immortal *Tetrahymena*
because these creatures had thousands of DNA strands and
therefore would be expected to have many copies of the molecular machinery that worked at the ends of DNA. At the time, there
were several competing theories regarding the origin of telomeric repeats. An attractive one was that the ends were spun by
a kind of molecular "spinning wheel," in this case made of
protein, that added the repeated DNA onto the very ends of the
DNA strands. This idea—first promoted by Elizabeth Blackburn, Janis Shampay, and Jack Szostak—proposed a hypotheti-
cal protein initially called "telomere terminal transferase."

Then, in 1984, Blackburn's lab performed a landmark
experiment. They took millions of DNA strands with the *Tetrahymena* telomeric sequence repeated four times—that is,
TTGGGGTTGGGGTTGGGGTTGGGG. They then put the
DNA strands in a tube with "mushed-up" *Tetrahymena*. The
resulting "cellular soup"—or "extract," as scientists prefer to call

it—would be expected to contain the telomere terminal transferase along, of course, with the thousands of other components
of the organism. The building blocks of DNA were then squirted into the tube—that is, the A, G, T, and so on—but the G
nucleotide was first made radioactive, thereby "labeling" it.
Assuming that telomere terminal transferase was in the mushed-
up extract, and that it "knew how" to add the DNA building
blocks onto the strands of DNA, one would expect that the
radioactive label would show up on the added telomeric DNA
strands.

This DNA was then exposed to X-ray film, as I'd done in
Sam Goldstein's lab (in fact, the same month I had my run-in
with Sam), and the results were visible on the film after developing. On Christmas Day in 1984, Carol Greider, then one of Dr.
Blackburn's students, went into the laboratory, peeled the X-ray
film from the gel, and ran it through the developing solution.
Holding the film up to the lights, she saw quite clearly and
remarkably a "footprint" of bands like a ladder of higher and
higher sizes extending from the original radioactive DNA.

Blackburn's lab had clearly shown that Olovnikov was right
in predicting that there was a special protein machine that could
manufacture the telomeric repeats. But all scientists could see
was the footprint of the protein. No one had found the actual
molecule that had made the footprints, the molecule called
telomere terminal transferase. But over the next few years, data
steadily accumulated in support of the theory that the molecule
existed. Because it had such a clumsy name, it was soon given the
simpler name *telomerase* (pronounced "tel-AHM-er-ace").

But were there any data out there supporting a role for
telomeres in cell mortality and immortality? I found a paper that
morning in the library, published in 1989 by Gregg Morin at Yale

University. Morin was interested in trying to find the footprint of telomerase in human cells. Because HeLa cells grow so aggressively, he mushed up a lot of these cells and reported in the journal *Cell* that he indeed had found the telomerase footprints. And, of course, it didn't escape my attention that this was in the *immortal* HeLa cells. However, Morin didn't say whether he did or did not find the same footprints in normal mortal cells.

These were all tantalizing clues, but footprints alone are not proof of a case. At four o'clock in the morning, bleary-eyed and light-headed, I turned the last page of an article in a book published in 1986 by Jim Watson's laboratory, reporting on the research of a scientist named Howard Cooke working in Edinburgh, Scotland. Again, not intending to test Olovnikov's theory, Cooke had studied a segment of DNA very near the end of one of the DNA strands. He observed that when he chopped up the DNA in precise pieces, the end piece, the one that included the telomere, always came out in a vast range of sizes—an ugly smear on a gel. Since DNA is normally copied quite accurately, and we all come from a single cell after the sperm and egg cells unite, the length of the telomere must have changed sometime after that, he reasoned. So he asked himself, "When did it change?" To answer the question, he measured the length of the telomere from a variety of subjects. The length of the ends of DNA, he found, were getting shorter with age in human beings. As he stated in his 1986 paper:

One possible mechanism for this loss could be that somatic cells, but not the germ line, are deficient in the type of terminal transferase activity demonstrated in tetrahymena, resulting in a loss of DNA at each cell division. This could imply that the number of terminal repeats would limit the number of all divisions possible.

But it was the figure summarizing his results that electrified me. I stared at the image before my eyes. His diagram of the immortal germ line, the cells with *zoë*, the cells with the green essence of life, was a flat line showing stable and long telomere size. But the somatic cells, after proliferating, were progressively losing telomere length, just as Olovnikov had predicted in his Russian publication back in 1971.

I had a problem. I was a captive medical student, it was morning, and I'd been up most of the night. But as I sat outside the library, having a cup of coffee before the slog of morning rounds, I believed that I'd just come across the very ticking clock of cellular aging. No, there was no proof of it, but I personally was convinced.

I stared at the rosy dawn of a new day. The cycling of the sun was both a promise and a curse. It was a promise that life renews itself in an immortal cycle just as the sun continues to rise. But it was also a curse, reminding me that one more day had passed and I had only a finite number of days left in my life to accomplish a terribly big agenda. I decided that I had to act, and I had to act now.

Chapter Five

Racing Against the Clock

But from the Fates' unbroken thread escape
Is none for those that feed on earth.
—PHANOCLES, quoted by Clement of Alexandria,
Stromata (Book VI, Chapter II)

Medical research is charged with tension. Like many areas of human experience, it pits the collegial, collaborative instincts most of us share—the sense of teamwork when we work together to accomplish a particular goal—against an intense, sometimes down-and-dirty competitiveness. On a good day, science is all about performing exciting experiments that keep you up all night; it is the open sharing of ideas and dreams over a beer in the village saloon. On a bad day, it is all about locked desk drawers and bickering about the order of authorship on a joint scientific publication.

In the late 1980s, the competition to decipher the clock of cellular aging gathered like storm clouds on the horizon. If one were to stare long enough at the scattered pieces of the puzzle, and in particular, five pieces that fit together quite nicely, one could see the unmistakable face of Alexy Olovnikov peering out from his Moscow subway station. The first piece of the puzzle was a confirmation of specialized structures, or telomeres, on the ends of human DNA strands in the nucleus. And the Woody Wright/Len Hayflick experiments showed us that the clock resided in the nucleus. Second, the shortening of telomeres provided a credible model of a "clock" that would allow a cell to "remember" how old it was, as predicted by the "Dirty Old Man Experiment." Third, Gregg Morin at Yale had found telomerase in HeLa's immortal cells; and fourth, Drs. Carol Greider of Jim Watson's Cold Spring Harbor lab and Bruce Futcher and Cal Harley at McMaster University in Hamilton, Ontario, saw telomere shortening during the aging of cells in vitro. Fifth (and for me the most definitive piece of the puzzle), Howard Cooke had observed that telomeres shortened in human somatic cells aging naturally in the body, but remained long and constant in length in cells cultured from the male germ line.

For detectives on the trail of the mystery of cellular aging, the eventual shortening of telomeres to a critically short length could therefore signal DNA damage to cells, damage severe enough to cause them to throw on the "emergency brakes" and stop dividing. This would explain why tumor viruses that disarmed these brakes might help to immortalize cells. It would also explain why super-old cells in mortality 2 would have fused telomeres.

Looking back at the development of our understanding of the clock of cell aging, Olovnikov's prescient prognostications seem somewhat surreal. How could a man with no data other than the fact that somatic cells displayed a finite life span conceive of the whole telomere hypothesis? It would be like someone hearing that a lost piece from Bach had been discovered, then walking over to a piano, sitting down, and playing the whole piece, never having heard or read a single note.

Nevertheless, I now believed that he was right—but I knew that I was way out on a limb of deductive reasoning. Like the *EcoRI* and *Alu* repeat theories, the telomere hypothesis could easily give way to some new data that would bring the whole line of supposition crashing down.

To test the theory, there was a mountain of work to do. New and more sensitive tools to detect the footprints of telomerase had to be invented. We had to test for the first time whether it was active in all immortal cells and, as we expected, absent from mortal ones. And we had to test for telomerase in actual human tissue. We also needed a simpler way of testing the telomerase hypothesis short of finding the gene. If we had the telomerase gene in our hands, then it would be a simple task to introduce it into mortal cells and see if it made them immortal. But we didn't have it, and we didn't know when we would.

I had to decide what to do. Yes, I could finish my M.D. degree, write grants, assemble maybe a dozen lab personnel if I was very lucky, and then start to work on the effort I had in mind. This is the orthodox way such projects are undertaken. Alternatively, I thought, I could try to launch a biotechnology company and eventually, over a three- to five-year period, end up employing perhaps 120 people—far more, needless to say, than would fit into the cramped quarters of a medical school lab.

Simple math suggested to me that a biotech strategy could accelerate the discovery of telomerase tenfold. Perhaps more important, if I could find enough capital, I could form alliances with leading university and nonprofit laboratories working on telomeres. I reasoned that while normally these labs saw one another as competitors, they could comfortably share their findings in a joint venture with a neutral party like a company. A commercial enterprise cares a good deal less about publication in scientific journals and competing for grant support than about developing and manufacturing medical therapeutics. I felt that the company, like the hub of a wheel, had the potential to fund and unite a far more powerful effort than could possibly be launched in the normal academic setting.

For some years, including the period at Baylor while I was working on my Ph.D., I had nursed along a small Michigan truck-leasing company. Converting it to biotech was no big deal; I'd already used it to purchase some lab equipment I needed to follow up on some of my crazy ideas that were too unconventional for my Ph.D. program. Late one night, working in the lab at Baylor, I hit on the name "Geron." I knew from my years of reading the New Testament in Greek that *geron* means "old man." It is the root of the word *gerontology,* which refers to the scientific study of aging.

Geron appears in the New Testament just once, in the gospel of John. When Nicodemus asks Jesus, "How can a man be born when he is old [for which the New Testament uses the word 'geron']? Surely he cannot enter a second time into his mother's womb to be born again?"

I wanted to symbolize my vision of a biotechnology company focused on gerontology through a respectful salute to my long-term study of the Bible.

My earliest work at Geron had focused on agents that extended the life of cells (agents I dubbed "senstatins," meaning that they put a temporary "stasis," or halt, to cell senescence). This kind of technology had great promise in the many manifestations of cell aging, but I was concerned that merely extending the life of cells without finding and rewinding the clock of aging itself would simply make super-old cells (less of a fountain of youth, in other words, than a fountain of old age). Telomerase, on the other hand, offered a means of not just extending the life span of cells, but immortalizing them. Like the key used to rewind antique clocks, it had the potential to rewind the clock of cell aging. So while still in the monstrously hectic third year of medical school, I set out to build the company, with a focus on telomere biology.

Woody Wright and his colleague at the medical school, Jerry Shay, were the first people I asked to join my circle of collaborators, called the Scientific Advisory Board. Though a bit surprised by the unconventional approach I took to medical school, their approach was "Mike, this sounds wild and crazy, especially for a medical student! But, hey, why not?" and they signed on. We huddled together to write the first patent that would lay the foundation for Geron's telomere programs.

I tried to imagine every use of the telomere, from inhibiting the protein to stop the growth of tumor cells, to extending telomeres to treat the problems of aging and diseases that exhaust cell life span, such as atherosclerosis, AIDS, skin aging, and macular degeneration (a cause of blindness in the aged). The U.S. Patent

Office was very generous, and some of the literally dozens of patents we applied for and obtained in those early years would prove valuable down the road. There were, of course, a few major problems in trying to launch the company. I couldn't actually prove that all of this was real—that *any* of it was real (not exactly a minor point when asking investors to cough up hundreds of thousands of dollars in seed money). Instead, I had to convince them to believe in my instincts. And I had only vague ideas of how we were going to track down the telomerase gene.

For all I knew, I was going to need tens of millions of dollars in cash before we were done. And whoever invested in the enterprise would have to understand clearly that Geron would burn through large sums of funding on an annual basis for ten years or more before it could have any significant sales. Where was I going to find such intrepid investors? And how could I convince them to back an admittedly young and inexperienced entrepreneur? I needed what founders refer to as "angels" (a term borrowed from Broadway, where angel investors are those who back a play long before it can show a return of any sort on their investment). This would give me the capital to prosecute my patents, set up a respectable office, and travel to meet the really deep-pocketed investors.

An old friend of mine named Robert Peabody, who lived in Dallas, was a CPA and could, I imagined, provide some assurance to investors that their money would be managed in a professional manner. Over hamburgers and fries, I laid out my vision for him and asked for his help.

Bob asked me any number of questions along the way, tough questions, some of which I couldn't answer. But when I was done, he tilted back in his chair and, in a feigned Texas drawl, answered, "Well, okay, ole buddy, let's rock and roll."

Over the following months, Bob Peabody and I traveled across Texas in our spare time, trying to get people to hand over their hard-earned cash while we quickly burned through our own. More than once, my friends lectured me about the wisdom of spending all my savings on a harebrained scheme that, let's face it, might never happen.

Maybe they were right, but I had to try. It was tough going, and I soon reached a point when my last best hope lay in a wealthy man living in California who hated his own aging and had underwritten similar projects in the past. He, I convinced myself, would see that I was the scientist he'd been looking for. After finally getting an audience with him, I journeyed to the sunny seaside community of Santa Barbara, California.

Once admitted to his offices, I anxiously assembled my slide carousel, turned down the lights, and launched into my well-prepared presentation.

I was some ten minutes into it when my heart stopped. Glancing away from the screen, I saw my potential investor's head droop. With a start, it lifted. Then it drooped again, and this time it didn't rise—not at all. He was stone cold asleep.

What does one do in such a moment? Should I continue as if he were listening? Or pause until he woke up and then kick in where I'd left off, pretending I hadn't stopped?

Or should I just pack up my slides and go back to medical school, where I belonged?

Utterly undone, I chose the last course.

It was dark and rainy back in Dallas as Bob and I sat around my kitchen table asking ourselves why in the world investors wouldn't bet on the greatest revolution in the history of biotechnology. So many people we'd talked to had wished us well. But all the wishes and promises in the world weren't going to help me pursue my vision. Maybe at the end of the day I was going to have to pursue it alone, using the slow, academic route.

Then, at the eleventh hour, we got a lead from an unexpected source.

Karl Riabowal, once my fellow student at Sam Goldstein's lab, had since moved on to Jim Watson's Cold Spring Harbor Laboratory on Long Island. He called to tell me of a friend named Bill Ryan, an M.D. who was now also a biotech analyst at Smith Barney. At his urging, Karl set up a meeting for us in New York, and Bob and I packed our bags for yet one more dog-and-pony show. This one, I explained to Bob, would have to be the last. I couldn't afford any more time away from medical school, and my checking account was bone dry.

Smith Barney's conference room overlooked midtown Manhattan in a towering building on the Avenue of the Americas. Once again I launched into my spiel, candidly sharing my enthusiasm, portraying the world of possibilities that awaited us when the science of gerontology was applied to modern medicine. At the conclusion of my talk, some forty-five minutes later, Bill Ryan gave me immediate feedback.

"You know, this could be the greatest revolution in the history of biotechnology!"

That was Bill Ryan speaking, not me.

A few weeks later, he flew to Dallas, and there, on the spot, he wrote us a check. In the lobby of his hotel, Bob Peabody and I stopped to stare at the slip of paper in my hand.

"Bob," I confided, "now we have to decide whether we will cash it."

We were Caesars at our Rubicon. Cashing that check meant committing ourselves. Once we cashed it, we were going to have to make the company work. Even in our most optimistic forecasts, that would take years. Among other things, it would eventually mean chucking my years of hard work at medical school out the window.

We cashed the check.

Having our first investor (especially an M.D. and a biotech analyst) gave comfort to others, and soon other checks began arriving by Federal Express. One of my favorite angels was a dear man named Miller Quarles, a retired petroleum geologist who invested with the admonition that we had to conquer the "old age disease" while he could still benefit.

In a relatively short time, I had $250,000 in Geron's checking account. I set up an office in a tall glass-and-steel office building across the street from the medical school and now wore three hats, those of a full-time third-year medical student, CEO of my own fledgling company, and a senior scientist in Woody and Jerry's lab. There wasn't a lot of time to sleep in those days. One sunny April afternoon in 1991, while I was studying in my office, the impossible happened. I'd just started nodding off when the phone jangled me awake. The caller introduced himself, then said he worked for Alan Walton, and that he was calling from Oxford Bioscience Partners.

Alan Walton was a well-known venture capitalist in our field and a principal of the firm. Up to that point we hadn't been able to attract the attention of the venture capitalists. The business plans we mailed ended up being thrown in with the stacks of thousands of other unsolicited proposals. This guy was telling me that Alan had traveled the world looking for a scientist to start a biotechnology company that focused exclusively on aging. Three different sources had led him to Geron.

On June 3, I flew to New York and met with Alan and other members of his firm, laying out my vision. Alan's enthusiasm was written in a broad smile. He introduced me to the Rocke-feller family's venture fund, called VenRock, and then to other venture capitalists. To my amazement, they were all interested in these wild ideas.

Oxford Bioscience Partners and the accounting firm of Ernst & Young sponsored an annual forum in Redwood Shores, California, that featured trends in biotechnology investing.

"Would you come and present Geron?" he asked.

I was scarcely prepared for what followed. On November 6, 1991, I flew to San Francisco, rented a car, and drove to the Hotel Sofitel, near the heart of Silicon Valley. I found myself at the National Conference on Biotechnology Ventures, a gathering of hundreds of venture capitalists and people in the industry. Alan himself rolled out the red carpet for me. "As the biotech industry matures," he told our audience, "the technical successes of the past are continuing to present intriguing new opportunities. . . . Modern cell and molecular biology research is beginning to unravel the fundamental mystery of how cells and, by extrapolation, whole animals age." He described Geron Corporation as the up-and-coming leader in the field. And finally, he introduced me.

I grabbed the microphone like a life raft, launching once more into my familiar presentation. But this time, I felt particularly energized. Passionately, I made the strongest case I could that we were facing one of the most profound demographic shifts of our time, the aging of our population. And how we now had a model of aging in the test tube, the aging of the human cell. I explained how, by using the same DNA technologies that had given us biotech companies like Genentech and Amgen, we could now understand the genetics of aging and discover interventions that would address the trillion-dollar marketplace of an aging population.

I concluded, with all the fervor I could muster, that we stood at the dawn of a new era in medicine, a day when fundamental discoveries about the genes controlling human aging could and would lead to dramatic advances. Ultimately, I said, we would find the genes that control the aging of cells.

Then Alan took the microphone from me and asked for questions.

A long and awkward silence followed.

Had they not heard me? Had they, too, like my onetime prospect in Santa Barbara, all fallen asleep?

But then, out of the darkness toward the rear of the auditorium, a lone croaking male voice called out:

"Hurry up!"

It brought down the house.

I had packed up my slides and was heading for the door of the auditorium, when suddenly I found myself surrounded by people wearing badges displaying the name Kleiner, Perkins, Caulfield, and Byers. Of course, I had heard of KPCB. They were considered by many to be the most prestigious venture firm in Silicon Valley. I pulled out some business cards to pass around, but then realized I'd misjudged their intent.

They didn't want to exchange cards—they wanted my soul. And they wanted it there and then! They tightened in a phalanx around me to keep other venture people at a distance. I felt like I was being arrested.

I have always been impressed by decisiveness, but I must say I had never seen, nor did I know how to respond to, this kind of high-gear salesmanship. They slid me into an adjoining room, where they held forth about all the entrepreneurial companies they had launched. The list was long and impressive: Genentech, Amazon.com, AOL.com, Compaq, and Sun Microsystems, among others. Finally, they asked me to commit then and there to allowing Kleiner Perkins to lead the financing for Geron.

My father had taught me a thing or two about business—all the commonsense rules of closing a deal. But the most important lesson of all was never to walk away from a good proposition the moment it was offered.

So I shook hands that day with KPCB.

A few months later, at a conference table in a Palo Alto law office, after twenty minutes and seemingly a hundred signatures, we closed the first $7.5 million round of funding and broke out the champagne. Back in Dallas, I packed my U-Haul and drove the long drive to the San Francisco Bay area. It was a sheer pleasure working with professionals who knew and understood the high-risk business end of biotechnology, and in the months that followed, I came to understand how good a choice I had made. Having an investment group that included such sophisticated investors lent credibility to a company that many in the biotech industry thought was a bit over the top. The word on the street was that Kleiner Perkins had started a new biotechnology firm to clone the immortality gene. Our presentations were always standing room only, but I sensed from the beginning that many industry analysts were waiting for the moment when Geron would be unmasked as one great con job.

Of course, I was acutely aware of their skepticism, and it didn't help that I was such an unknown. However, I believed that the right way to counter all our critics was simply to do some of the best science ever done in a biotech company, and with this in mind, I hit the road running in the spring of 1992.

To find the telomerase gene, we needed the very best minds in telomere biology, and we needed "gene jocks"—bench scientists skilled at manipulating DNA.

One of my first stops was Cold Spring Harbor on Long Island, the famous laboratory run by James Watson that employed Carol Greider, among others. Jim hardly ever lent his name to biotech enterprises, but he was strongly interested in telomeres and he understood the potential in what we were doing. "This won't come in time to help me personally," he dryly remarked, "but, yes, I'll give you a hand wherever I can." We struck an arrangement, and one of the first ways Jim helped us was by encouraging Carol in her effort to find the telomerase gene in the immortal pond-water animal *Tetrahymena*.

Jim was particularly keen that I convince Gregg Morin to join our collaboration. Morin was the Yale scientist who was the first to observe the footprints of telomerase in HeLa cells. He had since moved on to the University of California at Davis. I drove through the vineyards of Napa Valley toward Sacramento with the goal of convincing him to leave his lab to become an employee at Geron. After a few weeks of deliberation, he left his career in academia to join the race. Then I twisted the arms of Cal Harley, Gunter Blobel, Len Hayflick, Elizabeth Blackburn, and others to become part of the collaboration in one capacity or another.

Len Hayflick had retired from active research and had put his entire laboratory in storage, but he really had no use for the aging stockpiles of equipment. I was thrilled when he offered to ship a large truckload of laboratory supplies to my empty facility. Launching Geron's labs with Hayflick's equipment imparted a sense of history and continuity. Every day, every experiment would remind me of the giants upon whose shoulders I now stood.

But we needed new tools as well, ones more sophisticated than those available in Len's day. For example, we needed a more sensitive test to determine whether there was active telomerase in a given cell. The method Gregg Morin had used to detect telomerase activity in HeLa cells was terribly insensitive. It just wouldn't do for looking at things like actual human tissue, where only a tiny fraction of the cells might be immortal. While I was still in Dallas, Woody Wright, Jerry Shay, and I had filed a patent on a technique we later called "TRAP," an acronym for Telomeric Repeat Amplification Protocol. This technique used "PCR" as a way of making billions of copies of DNA, starting with only a few copies. PCR is best known for its use in police work to identify who left a small spot of blood, and it played a prominent role in the O. J. Simpson trial. We adapted the technique to amplify the footprints of telomerase. I hoped that we would be able for the first time to peer into normal cells and cancer cells and determine whether Olovnikov's theory really was consistent with observed facts. The appearance of a microscopic quantity of the TTAGGG repeats on the end of an artificial telomere would provide the forensic evidence we needed to pronounce a verdict on whether telomerase was present at the scene of the crime.

Of course, there had been talk for many years about immortalization as a step in the transformation of a normal cell into a cancer cell. This largely reflected the observation that tumor cells grown in the laboratory dish, such as the famous HeLa cells, did not undergo Hayflick's senescence. But then again, had anyone really looked? No one actually knew if immortalization occurred before or after the tumor cells were grown in a laboratory. So no one knew if immortalization occurred in the tumor itself or was merely a laboratory curiosity.

So we began with what are called tumor cell lines, immortal

cells derived from human tumors, and took in all the lines we could get our hands on. I called the National Cancer Institute and asked them how many they could give me and then bought them all. Our incubators at Geron were soon overflowing with dishes, and many times angry technicians complained to the other members of the management team about my monopoly on the incubator space. I figured we needed forty large dishes (about 5 inches in diameter) for each of more than a hundred cell types. Our facility on Constitution Drive in Menlo Park, California, was a one-story modern biotech facility with plenty of lab space, but the lab outfitted for cell culture, with the blinking incubators and sterile hoods that I was used to working in from my days with Sam Goldstein, was in fact quite small.

Dishes of tumor cells soon metastasized to every available inch of the incubators—dishes of melanoma, ductal breast cancer, bone and prostate cancer—while a Geron scientist, Nam Kim, performed TRAP assays on these cultures, painstakingly performing experiment after experiment. The results began to trickle in, and they were wonderful, like a dream. We would eventually conclude that in the case of skin cancer cells, eight of the tumor cell lines were telomerase positive, while in none of the five normal skin cultures was any telomerase detected. The same was true with breast cancer: All of the cancer lines were positive, none of the normal breast cells. The studies went on and on. We looked at cervix, kidney, prostate, pancreas, lung—cells from a total of eighteen different tissues. The grand total: 98 of 100 tumor cell lines were telomerase positive, and none of the 22 mortal normal cell cultures were.

These were important findings. Ever since President Nixon had launched his "war on cancer" in 1971, there had been growing public frustration that science wasn't delivering a cure in the face of the million new cases occurring every year. In response, an ever-increasing number of scientists were chiming in that there was no such thing as "cancer." Instead there were "cancers"—cancer of the kidney, totally distinct from cancer of the lung, and so on. There was no unifying theme to all of cancer, they explained. "So give us more time and give us more money," they pleaded. "We have to study each cancer one at a time."

The closest thing to a common molecular pathway to the cancers at the time we conducted our TRAP experiments was that up to half of all cancers had mutations in a gene called "p53." The p53 gene functioned as a kind of emergency brake used by a normal cell to control its own growth. In about half of all tumor cells this brake was broken, an important step to becoming an out-of-control cell. But no one had come close to finding a cellular alteration that was found in a majority of all types of cancer or that pointed toward a new therapy. Perhaps we had found such a universal or nearly universal change in cancer. And then again—who could say?—perhaps it was a universal marker for cancer cells grown in the laboratory dish but would prove useless in real people.

So we needed to look at actual tumors. With Jerry Shay's help, we started collecting malignant tissues and their normal counterparts as controls. I still remember the first call from Jerry. He had measured telomerase in three malignant breast tissue samples and in two normal samples of breast tissue. All three of the cancer tissues were positive, and both the normal tissues were negative. Two weeks later, he had more breast tissue samples and some prostate tissue as well. Over the following weeks, the excitement mounted. At the end of the study we had surveyed several hundred mortal and immortal cell lines, and almost as many from malignant tumors, and compared them with their normal counterparts. The results were almost too good to be believed.

We had looked at a total of twenty breast cancers that were relatively advanced, and eighteen out of twenty were telomerase positive. When we then turned our attention to normal breast tissue, not one in eight tests showed the telltale activity of the immortalizing gene. Germ-line tissues such as testis and ovary were universally telomerase positive. We finally extended the studies to colon, head and neck, prostate, brain, lung, uterus, blood—in total, 12 human tumor types and 50 normal somatic tissues. In the final tally, 90 out of 101 tumors were telomerase positive, and none of the 50 normal somatic tissues were. Even this modest failure on the telomerase-positive side was predictable, because we knew that some of the tumors had likely degenerated; in addition, the TRAP assay didn't always work. Nonetheless, we had a correlation of telomerase activity with cancer that was over 90 percent!

It's hard to find words to communicate the excitement of scientific discovery followed by its repeated confirmation, knowing that your eyes are seeing for the first time one hidden piece of the mystery of life. This is especially true when it involves a problem as significant as cancer. I suppose we have all known someone who has suffered a slow and painful death from this scourge on mankind. In my life, my aunt lived with our family as she was dying of lung cancer. I will always remember the sounds of her coughing and sighing late at night from the intractable pain.

I called the National Cancer Institute and informed them of our striking results. Because telomeres were an unknown factor in cancer research at the time, a meeting was organized at the NCI so I could get together with the director, Sam Broder, and his staff. Carol Greider offered a brief background on the biology of telomeres, then I presented the data we had collected from cells and tumors. When we were done, Sam Broder leaned back in his chair and uttered the most complimentary one-word summation a contemporary scientist gives to another: "Cool!" We all laughed.

We published our results on telomerase and cancer on December 23, 1994, in the journal *Science*, almost ten years to the day from the time Blackburn's laboratory first saw telomerase activity in the immortal pond-water organism *Tetrahymena*. As is so often the case, the critics—and we had plenty of them—responded resoundingly. "Correlations! Correlations! Correlations!" demanded my old colleagues. Where was the hard evidence of causality? All we had shown was that immortal cells *had* telomerase, not that telomerase was actually *causing* the immortality. Therefore, the naysayers opined, our studies were really of little value. After all, there are hundreds of changes that take place in cancer cells; this was merely one more.

Still, we believed that we had uncovered what could potentially be a wonderful new target for cancer therapy. The dream of cancer researchers has always been to find a protein that is present only in tumors, present in all types of cancer, necessary for the growth of the cancer cell, and a molecule that could, in principle, be inhibited by a chemical, a pill. Telomerase appeared to be such a candidate. While it was turned on in the reproductive germ-line cells and quite possibly certain other cells in the body, for the most part tissues had no active telomerase. If a drug could be made to inhibit telomerase and no other proteins, then we predicted that the tumor would promptly begin to age and would die, leaving the rest of the body largely untouched. This would be far superior to the "cut, burn, and poison" strategies now utilized and euphemistically called "surgery, radiation therapy, and chemotherapy."

All this talk of cancer, and the press reports that followed our paper on the first common link in nearly all cancers, caused many critics to complain that Geron was merely a cancer company. "We knew all this talk about anti-aging was just hype," they said. They complained that we had never shown any evidence that telomeres played a role in the aging of cells or that we could invent a means to intervene in aging itself. I didn't share their dim view of the telomere–aging connection; but I *was* concerned about how many hours in the day I found myself working on cancer, not on aging. We just *had* to find the telomerase gene. I placed much of my hope for getting the gene on one scientist, Carol Greider.

Carol and a student named Kathy Collins had painstakingly grown liters of the microscopic swimming *Tetrahymena*, mushed them into a fine paste, and separated the resulting proteins through tall glass columns of a gel-like substance. It took many hours of standing, shivering, in a refrigerated room, and many months of backbreaking work, but finally Carol sent us the detailed sequence of two purified proteins that she said were the two proteins that made up the enzyme (at least in *Tetrahymena*). With these DNA sequences in hand, we began a diligent search for their human counterparts. We ended up spending several hundreds of thousands of dollars in the process, but we consistently drew a blank. A new recruit to Geron, Karen Chapman, could find the counterparts to Carol's genes in related species of *Tetrahymena*, but a zoo-blot (a filter with DNA from a wide array of species from fish to amphibians and humans) did not react with Carol's gene.

On June 2, 1995, Carol published a paper in the journal *Cell* titled "Purification of *Tetrahymena* Telomerase and Cloning of Genes Encoding the Two Protein Components of the Enzyme."

In this report, she disclosed the secret data on which we had, by then, invested millions of dollars and thousands of hours of internal research, trying to find the human counterpart.

And consistently failing.

But why?

Although she published this with our permission, Carol's publication put us in an awkward predicament, because it meant that other scientists could now join freely in the hunt, and our backers inevitably started to second-guess us. How many years were they going to underwrite our failures? How did we know the scientists we'd recruited at Geron really were first-rate? Did we perhaps need some "expert" advice? In the end, we were cajoled and coerced into bringing in leading molecular biologists like David Botstein of Stanford University to look over our shoulders. But all they could do was tell us what we already knew. After all, we ate, drank, and slept with our research.

One possibility we considered was that the human and *Tetrahymena* genes were just too different to provide a useful comparison. Maybe you couldn't "walk the evolutionary tree" all the way forward to the human species. After all, human beings don't look a heck of a lot like the billion-year-old residents of pond scum.

But my bias was that genes with ancient roles like this usually don't change that dramatically. For example, consider the fact that the sequence of DNA made by *Tetrahymena* telomerase was TTGGGG, only one base different from the human counterpart of TTAGGG, even though the animals split on the evolutionary tree more than a billion years ago.

Indeed, we were gambling heavily on the fact that there should be such similarities. One of our technicians, Junli Feng, had been working hard trying to fish out the RNA component of telomerase from human cells based on the assumption that they were similar. If they were not, then she was wasting the better part of a year of hard work.

We expected that telomerase had at least two pieces. Like the Fates that spun the thread of life, it was expected to have an active component that actually did the "spinning"—that is, actually made the telomeric TTAGGG. That was the "hand" of the enzyme. And telomerase was expected to have another piece, like the ruler laid against the thread of life by Lachesis. This ruler was expected to be a ribbon of RNA that instructed the hand to make "TTAGGG" as opposed to some random sequence like "GAGGCG."

If the above were true, then telomerase would be in the class of molecules called *reverse transcriptases.* A reverse transcriptase is a molecule, normally found only in entities like viruses, that makes DNA starting with instructions in an RNA molecule. No one had ever found a reverse transcriptase that was a natural part of a human cell—that is, not associated with a foreign virus. But we believed that telomerase should be there all the same, acting like the viral reverse transcriptases, with at least two pieces, an RNA piece and a protein component.

One day Junli called me over to her bench.

"Mike, I have a strong band here on this gel, and it is about the size of *Tetrahymena* telomerase RNA. There are some other bands here as well."

She showed me a band on a gel. It represented a small human RNA, and it was fished out because of assumptions that it was like the *Tetrahymena* RNA. I somehow knew I was looking for the first time at the first piece of the immortalizing enzyme.

"That's telomerase," I said, as if saying it confidently would make it so. I guess it was the strength of the band, and the fact

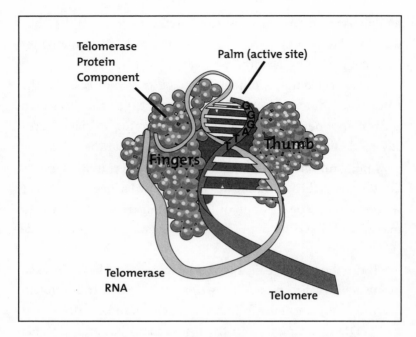

Figure 10. The Twin Components of the Immortality Enzyme. Like the hand of Clotho, the enzyme was expected to bind to the telomere and synthesize the telomeric repeat sequence "TTAGGG." The code for the repeat sequence was believed to reside in a ribbon of RNA.

that it had the approximate size expected of the telomerase RNA component.

But we had to know for sure we had the right RNA, and we were hearing that we had fierce competition. So I devised a quick test. I had earlier prepared a library of RNAs from the normal mortal cells and immortal tumor cells we had previously used in our cancer studies. We also had RNA prepared from testicles, which we expected would be rich in the immortalizing enzyme. I took the RNAs and coded them in my lab book with numbers and handed Junli the tubes numbered only 1–10. In reality, 1–5 were mortal cells, 6–10 were immortal tumor cells. I then left for a fund-raising trip to New York. I received the eagerly awaited call at 6:30 P.M. It was from a colleague, Bryant Villeponteau.

"Well, the first five tubes were mortal cells, because we see bands but they're dim. The next five samples were the immortal cells, because the bands are much brighter, and the last tube you gave us is from the testicles, because the band there is the brightest of all."

I hung up the phone confident that we had the right gene. I brought a bottle of champagne to the room of a colleague to celebrate. We were at least *part* of the way there. We had the first piece of the immortalizing enzyme, and we hoped that would give us an important patent position.

Perhaps most important, we could now perform an experiment in which, using some molecular trickery, we deleted the RNA from the immortal HeLa cells. The question we would ask was whether over the following weeks the HeLa cells' telomeres would shorten, and whether eventually the cells would die. In the following days, the excitement grew as we learned that telomeres were indeed shortening. And most breathtaking of all, the normally immortal tumor cells were now aging and dying! This and the other data we subsequently published in a scientific paper convinced us and all the critics in the scientific community that the first piece of human telomerase had been cloned. But there was still another side of the coin. We had shown only that by eliminating telomerase from an immortal cell, we could convert an immortal tumor cell to a mortal one. We had not yet shown the converse—that adding telomerase to a mortal cell would make it immortal. It was, to be sure, the strongest evidence yet that we were on the right track. But it was not proof.

That weekend I drove the long coastal road to northern Marin County to spend the weekend with Len Hayflick and his wife, Ruth. They invited me to dinner at an elegant restaurant overlooking the Pacific Ocean. Gazing out over the surf, I told Len of my line of logic leading to telomeres as the clock of cell aging, and then I dropped the bombshell.

"Len, we get telomere shortening and senescence in HeLa cells when we eliminated the RNA component. The good news is that it looks like telomeres really are the clock of cellular aging."

Len leaned back in his chair, turned toward the ocean and the setting sun. He had published papers on the complexities of cellular aging, pointing out the many hundreds of changes seen in aging cells, data that for many researchers lent more support to the "wear-and-tear" models than genetically regulated senescence.

"God, can it really be that simple?" he asked out loud.

But the critical piece of the puzzle, the piece on which all the other pieces hung, the key to the vindication of the telomere hypothesis and our survival as a company, was to get the active protein component of telomerase—the "hand" of the enzyme. We essentially didn't have anything unless we had that gene. On that long drive home along the Pacific coastline in Marin County, the conviction grew within me that we just *had* to win the race in getting that gene.

Meantime, beginning in 1993, another ambitious researcher had entered the fray. Joachim Lingner was a student of the Nobel Prize–winning scientist Thomas Cech at the University of Colorado at Boulder. Like Carol, Joachim had the idea of isolating telomerase from an immortal pond-water animal, but he chose the organism called *Euplotes,* which had an astronomical 20 million chromosomes, or 40 million telomeres. In theory, this meant that there would be at least 40 million copies of the telomerase protein in each cell.

Still, the task Joachim and his collaborator, Loren Denton, gave themselves was Herculean. They, like the scientists in Carol's lab, would need to grow hundreds of liters of *Euplotes* in five-gallon flasks and learn to keep them happy, feeding them the algae they needed in order to thrive and reproduce. After thousands of hours of diligent effort (and many billions of sacrificed *Euplotes*), Joachim succeeded, in 1996, in purifying the protein he sought. He sent off a tiny bit of it to a specialist in sequencing scant traces of protein, a German scientist named Matthias Mann at the European Molecular Biology Laboratory. Matthias had developed a state-of-the-art technique, called "nanoelectrospray tandem mass spectrometry," that allowed him to read off the sequence of the protein. Working backward from Matthias's sequencing, Joachim was able to amplify the DNA sequence in *Euplotes*. Then came an explosive discovery. As he entered the sequence into his computer, he observed that the protein he called p123 had a striking similarity to retroviral reverse transcriptase, a protein in the AIDS virus that makes DNA starting with RNA. Furthermore, he discovered that his sequence bore no similarity whatsoever to the one Carol Greider had published.

Joachim came to Geron and presented his data. On the way back from his talk, Karen Chapman and I chatted about the results. She was impressed by the careful and precise way in which he had shown how his p123 protein tracked with the activity of the enzyme, and we were both impressed by the relationship to reverse transcriptases. By the time we reached our offices, we were both thinking the unthinkable: Carol may have isolated a protein associated with telomerase, but it clearly was not the critical component. It meant that we had spent many precious months and dollars on the trail of the wrong suspect. We now had to massively redirect our research efforts. As far as we knew, no one had yet found the human gene. That was all that mattered. But who could tell how long that would last?

To top it all, the biotechnology giant Amgen, which could have easily swallowed Geron, announced in its 1997 annual report that it had cloned one of the components of human telomerase! Some of our investors panicked at the news. All Amgen had found, we realized, was, at best, a protein associated with telomerase, the human gene corresponding to the one Carol Greider had described. They, too, had started from the data in her paper. Once again we had to convince our doubters that there was no need to panic. "Trust us."

Immediately upon filing the patent for Joachim's work, and under a strict confidentiality agreement, the University of Colorado gave us a peek at what they had found. In parallel, they sent the same information to several large pharmaceutical companies. Thankfully, we were the only ones to take the bait. Tom Cech said later, "One of the Big Pharmas said that if it had been the human [component], they'd have been very interested." But from a single-celled pond-water animal? As Tom put it, "They had trouble convincing their businesspeople it would be relevant."

By the skin of our teeth, little Geron got the exclusive license to the University of Colorado patent. But unlike the long period of secret access we'd had with Carol's data, Tom and Joachim, like most academic researchers, were eager to publish theirs before they got "scooped." Their paper, submitted to *Science*, would be published on April 25, 1997. At that point the whole world would have the same data. If we hadn't found the answers by then, we'd be back at the starting gate and on an equal footing with everybody else in the race.

I gave stern orders to our molecular biologists to watch the

Internet like a hawk. The Human Genome Project was by then in full swing, with pieces of it being published along the way and labs all over the world posting their discoveries as they came to light. If a piece of human DNA were found similar to Joachim's gene, that would give us the first piece of human telomerase, and almost certainly the definitive patent. They should watch the Internet daily for the first sign of a match in a human gene.

Then, late one night—it was around two in the morning—Toru Nakamura, another graduate student in the Cech lab in Colorado, was sitting at his computer. While waiting for an experiment to run its course, he dialed into the National Center for Biotechnology Information expressed sequence tag database. He typed in the sequence of the *Euplotes* telomerase gene and hit the Search key, asking the computer to look through the billions of human DNA bases for anything that even remotely resembled *Euplotes* telomerase.

And moments later, Toru struck gold. There, right before him on the screen, was a human sequence! It was a snippet of human DNA, of unknown function, posted by the University of Washington at St. Louis Medical School. It came from an inflamed human tonsil.

The next morning when he shared his find with Tom Cech, Tom confirmed that parts of the human DNA bore points of marked similarity to Joachim's gene and the other reverse transcriptases. Most important, the closest similarities were in the reverse transcriptase motifs. And it was these same regions where the similarity was found with the EST-2 gene, the gene that led to ever-shortening telomeres in yeast. This suggested that the human gene was human telomerase and EST-2 was yeast telomerase.

What was the probability that all this was mere coincidence?

When I heard the news, though, I was furious. Thank goodness it was our colleagues at Colorado who'd come up with the discovery and not someone else, but what on earth was going on inside Geron? It turned out that the human sequence Toru Nakamura had spotted had been freely available on the Internet for two weeks. And our scientists had been asked to watch the Genome Project on a daily basis.

How had we missed it?

It turned out that our people had been searching an outdated database by mistake, just the kind of fumble that might have cost us the game. Who was to say somebody else didn't have a two-week head start on us?

Tom Cech had faxed us the first look at the gene sequence—actually a small stretch of the gene we had searched for so long to find. The bottom of the page recorded its history. The small piece of the needle in the haystack began its life with the provisional name "Homo sapiens cDNA clone 712562." Once the sequence of the 389 nucleotides was deciphered (that is, once the A, G, T, and Cs were put in the right order) and it was determined to be a new and unknown gene, it had been lined up in the queue with many other unknown genes, like orphans hoping to be adopted by some loving parents, and given its second provisionary name, "Genbank accession #AA281296."

We were going to adopt the gene, decipher its entire sequence, and give it the name it deserved: "Telomerase," or more precisely, the catalytic component of human telomerase, referred to by most scientists as hTERT.

But then the bombshell blew up in our midst.

One of our prime rivals all along had been the illustrious Robert Weinberg, an aggressive, highly competitive, and brilliant scientist at MIT. Bob had been a leader in cancer research for many years, being among the first to characterize the broken genes that change happy normal cells into the runaway cells we call cancer. He had been a leader in the characterization of the broken accelerator of the cell that sends the signal to divide as fast as it can, genes called *oncogenes*. He had then played an important part in discovering the role of the broken braking mechanisms of cells, mechanisms that normally put a halt to runaway growth (genes called *tumor suppressor genes*). Now, with great resources behind him and with the same vision of Nobel Prizes dancing in his imagination that at once tantalized and propelled all of us, he too was in the hunt for the mechanism of the infinite fuel supply of cancer cells, the immortalizing gene telomerase.

The rumor reached Geron, and with it a pandemic of panic. At a conference in San Francisco, Weinberg had shown a short sequence of DNA that he claimed was a piece of the telomerase gene!

Immediately our conference room was transformed into a war room. If Weinberg was sufficiently confident to show the fragment publicly, he must surely have the rest of it too, or at least be very close. If not, why would he run the risk of giving potential competitors a leg up by giving them this free hint? We were certain of it, and that very certainty made us all feel a bit sick.

We'd already been operating under the assumption that the sequence we'd gotten from Cech was the real thing. The same day, Karen Chapman had compared its expression in the mortal and immortal cells we had collected previously, and she had shown me the striking result that it was strongly expressed by every immortal cell, and absolutely absent from every mortal cell. There was no way this could be mere coincidence. We had our hands on the real thing, but it may have been too late.

And so, at an emergency conference, we organized an all-out effort to sequence the whole gene as fast as humanly possible. I remember looking around the table; every eye was already bloodshot from lack of sleep. We couldn't run any faster. We'd been on a crash program around the clock ever since Tom Cech had sent us the sequence. But this was the life and death of our company. So right then and there, any planned vacations were canceled and cots were set up in a construction trailer in the parking lot. Commuting time was eliminated. People were literally going to live at the company, working every waking hour.

It took us about two weeks. In the summer of 1997, we wrote up a paper for the journal *Science* to publish our results.

Then the nail biting started.

Every week we waited to see if Weinberg's lab would scoop us by publishing first. Then the rumor arrived that he had indeed written a paper and submitted it to another journal, *Cell.* By then we knew ours was scheduled for the August 15, 1997, issue of *Science*. I was particularly concerned that Weinberg was at MIT, where *Cell* was published. He could likely get his paper out faster than we could. After all, he could just walk across campus and put a little friendly pressure on the editors. There was a strong chance we'd be scooped. We waited and watched every new issue of *Cell* with bated breath.

Finally, at 5:22 P.M. on August 7, 1997, Cal Harley sent out an office memo:

Subject: Relief.

The contents of the Aug. 8 issue of Cell just came online.
No Weinberg.

A week later our paper, titled "Telomerase Catalytic Subunit Homologs from Fission Yeast and Human," was published in *Science*. It created a terrific flurry of publicity.

I was in New York City the day the stories broke. My office faxed me Weinberg's paper, which was slated to be published the following week but which had now been released to the media early because of our paper. I quickly scanned his data and handed the paper to Karen Chapman. She sighed a sigh of relief, pointing out that their results so closely matched hers. Clearly, we'd gotten the details right. I rushed off to appear on CNBC and explain why our stock was up 115 percent that day and answer the big question: "What, after all, is all this talk about the 'immortality' gene?"

MIT sent out their own press release:

(Cambridge, MA) The discovery of a key molecule linked to the immortalization of human tumor cells provides an important new target for anti-cancer drug design. Researchers led by Dr. Robert A. Weinberg of the White- head Institute for Biomedical Research have isolated and cloned the gene for the long-sought catalytic subunit of human telomerase . . .

Weinberg's paper was officially published seven days later in the journal *Cell*.

We'd won.

To this day, though, I am amazed that after five years of intense effort, and millions of dollars expended in researching telomerase, we still ended up neck and neck in the homestretch. But we'd won indeed, and the banner that festooned our cele- bration at Geron read: "We won our Telome*race*!"

In retrospect, I think this was the first time Geron was ever treated seriously by the financial community. The fact that the Nobel laureate Tom Cech lent his name to our paper and that Robert Weinberg was also touting the importance of the gene and its role in cellular immortality left few doubting that we had done what we said we would do—and that it was important.

The world's press was quick to pick up on the scientific use of the word "immortality." *Time* magazine titled its coverage "The Immortality Enzyme." Bloomberg Business News covered the story as "Geron Stock Soars as Firm Clones 'Immortality' Gene." The *San Jose Mercury News* wrote a story titled "Endless Youth?" full of speculation about the extension of human life. But there was just one problem with all this hyperbole: No one had actually shown that the introduction of this gene was enough to get telomerase activity. Or what would happen if it were.

We obtained patent protection for the entire gene, which we later named "hTERT"—for *h*uman *t*elomerase *r*everse *t*ranscriptase. In this respect, the winner of the race—Geron Corporation—was decided by a patent examiner in Washington, D.C. We now owned the exclusive rights to the gene, and we could perform the crucial test of the whole telomere hypothesis.

Of the twin components of telomerase, we already knew the RNA component was ubiquitous in both mortal and immortal cells. Therefore, we reasoned, we had no need to add it to mortal cells. But the catalytic component, the gene called hTERT, was expressed only in immortal cells and was undetectable in mortal. Therefore, what would happen if we added it to mortal telomerase-negative cells?

The first "quick and dirty" experiment we designed was simply to introduce the hTERT gene into some mortal cells, then use the TRAP assay to see if any active telomerase had been generated. For all we knew at the time, telomerase could still have dozens of other components, in which case, when we introduced hTERT into mortal cells, nothing would happen. This would not be good, but frankly, it is what we had every reason to expect.

Scott Weinrich, who had previously led the effort within Geron to isolate telomerase from human tumor cells, performed the test in our lab, expressing hTERT in mortal cells, TRAPping them, then inspecting to see if there were any bands of a gel, evidence that we had awakened the immortalizing enzyme. This was still the summer of 1997, and Scott had developed the first gel. The word spread like wildfire around the company. There was telomerase activity in the cells after adding hTERT!

But was hTERT really all there was to it? Science is usually a long series of failures with intermittent success, just enough to lure you back to the laboratory. The result seemed just too good to be true. But the footprints of telomerase were indisputable. We had reawakened the enzyme.

Now it was back to where I started—the tissue culture lab, the incubators with the flashing lights, the plastic dishes of cells with the bloodred culture medium. All we needed to do now was "slip" the isolated gene into our cultures (literally dropping a small drop of liquid containing billions of copies of the telomerase gene in with the cells and giving them a brief electrical shock to push the DNA into the cell). We had modified the telomerase gene in such a way that the "switch" that determines whether it would be on or off could function only in the "on" position. So this all-important experiment would add telomerase to a mortal, telomerase-negative cell and then grow it through its life span. The question we were going to answer, of course, was: Would the addition of telomerase indeed stop the aging of mortal somatic cells? Had we therefore really found the immortality gene?

I thought of my old friends Woody Wright and Jerry Shay. They were extremely well equipped to help us in a project like this. Jerry often spent whole afternoons in the tissue culture suite in Dallas putting genes into cells, and Woody was a very competitive scientist, especially in a once-in-a-lifetime experiment. And since the days of the tumor studies, I had heard less and less from them. I missed our give-and-take.

I gave Woody a call.

"Woody," I said, "I have some good news. I am sure that we have telomerase cloned."

Woody was skeptical.

"How do you know it's the real thing?" he asked.

"It is expressed in all the immortal lines we've studied and is absolutely absent from mortal somatic cells," I explained. "But what makes me sure is that it has reverse transcriptase motifs, and even shares them with EST-2 [the probable yeast telomerase gene]."

There was a long pause. I knew Woody well enough to know he was thinking about what to do next, jumping ahead in his mind, clearing his mental decks of all the work-in-progress in his lab.

"What do you say we collaborate to test it on human cells?" I asked just to break the silence and make sure he was still on the line.

"I'm thinking we should use BJs," came his answer.

"BJ" was a cell culture that had been obtained from newborn human foreskin when I was still in Jim Smith's lab. BJ cells showed an incredibly long replicative life span, often ninety or more doublings before senescence took over. They would give us plenty of doublings so we could carefully measure telomere length and collect life-span data. But we needed to study more than just skin cells. In another collaboration, with Larry Hjelmeland of the University of California at Davis, I had been culturing retinal cells from the eye, those known as *retinal pigment epithelial cells* (RPEs). We agreed to use these and a third type too, *vascular endothelial cells,* ones that line the inside of blood vessels. All three of these cell types had been implicated in specific manifestations of aging—the aging of the fibroblasts in skin aging, RPE cells in age-related macular degeneration, and vascular endothelial cells in perhaps the most critical manifestations of aging, such as atherosclerosis, heart failure, and stroke.

But I suggested we first use Len Hayflick's cells.

One day at Geron, a Canadian film crew had been filming Len, who was visiting, and they asked him to demonstrate tissue culture. As gruesome as it may sound, Len hoisted up his trousers with the camera running, swabbed the calf of his leg, and, scalpel in hand, sliced off a small piece of skin. He later handed the tissue sample to me. I already had Len Hayflick's tissue culture hood and his microscope; now I had his cells, growing in the incubator he'd given me. Could anyone now seriously question his generosity? As I joked to my friends, now we could measure the *true* Hayflick limit! But didn't it seem appropriate that his cells would be the first human cells in history to meet telomerase?

Woody resisted my sentimental notions. He had his mind on more practical matters and thought BJ cells were far better suited to a well-constructed study. And so we went with BJs, as far as the definitive tests were concerned. But I was determined to introduce Hayflick to telomerase at Geron.

At first our group had a hard time getting the DNA stably integrated into human cells. Andrea Bodnar, a skillful scientist, suggested that something might be wrong with the telomerase gene itself. I didn't buy that. I couldn't explain why the experiment was not working, for normally getting DNA into cells is bread-and-butter cell culture. In frustration, I rounded up the supplies myself, walked into the lab, sat at Len's sterile hood, and uncapped the tube containing the telomerase gene. As I thumped the plastic tip on the handheld pipette used to measure out small quantities of liquid and dipped it into the DNA solution, I stopped for just a moment, thinking briefly of another day, some fifteen years earlier, when I had sat looking at senescent cells in Sam's lab.

Fifteen years! It had taken me fifteen years to get to this point. And now I held what just might be the immortality gene in my right hand, a gene that, all in all, had cost us tens of millions of dollars to isolate. And I had Leonard Hayflick's skin cells in my left.

I glanced at the clock. It was 6:38 in the evening.

I dropped in the DNA, placed the plastic vial containing the cells in a little device designed to zap DNA into cells, and pressed the button. After a second, it beeped. Then I sucked the cells back out of the plastic tube, reintroduced them into a normal tissue culture dish, and went back to my office.

Once there, I opened my notebook and recorded the work I had just done under the heading "The Immortalization of Dr. Hayflick." I closed the notebook and looked out of my window, staring across the parking lot at the building that held the tissue culture labs. That was that. I had good feelings about the experiment, and to celebrate, I headed into San Francisco for dinner with Karen. It was a beautiful night in North Beach, and Karen and I buried our emotions in wine and conversation, and the lights of the section of town known as Little Italy.

It took four weeks. Andrea was continuing to try to get the gene into some other cells and was testing the outcome by using our TRAP assay. I asked her now to test Len's cells for the telomerase footprint.

A few days later, she walked into my office.

"Congratulations, Mike," she said with a smile. "Your culture is positive."

Like a new father, I ran over to the tissue culture area to take a closer look at the cells. They were still proliferating, not showing any signs of senescence.

It was only a beginning. One culture wouldn't be enough for a paper, and Woody and Andrea were carrying out the much larger study where dozens of independent experiments were performed in parallel. Still, Len Hayflick's skin cells had been "telomerized," and I began to chart the course of their aging with and without telomerase on my office door.

In the following weeks, I posted an updated graph every three days, showing the growth curve of a control group of Len's cells without telomerase against the curve for those with the gene expressed. Within two months, the two curves, which had been neck and neck up to that point, started to separate. Len's untreated cells had begun to slow down. They had grown only twenty doublings, and then they stopped. But the telomerased cells continued to double, unabated, as though they were still young! Every few days, as I added a new data point, I believed a little more what many people might find unthinkable—we had halted the aging process of human cells!

On Friday, November 21, I cryopreserved Len's telomerased cells and left for a vacation (actually for a honeymoon, in Fiji, with Karen, my new bride). When I came home, Andrea's and Woody's data had already been written up and sent off to *Science*.

Our paper was published on January 16, 1998, but somehow it was leaked to the media three days early. We noticed the sudden rush of buyers of Geron shares and called Nasdaq to order a halt to the trading. We scrambled to finalize our press release and sent it over the business wire. "Extension of Human Cell Life-span Reported in Science: Telomerase Rewinds the Clock of Cell Aging," it cautiously reported. Our announcement caused quite a stir in the media and the scientific community, but nothing like the press release put out by Woody and Jerry at the Dallas medical school. It read, "By all accounts these cells had found their cellular fountain of youth," and "The potential long-term applications are simply staggering."

In spite of my displeasure over the use of the term "fountain of youth," I couldn't disagree with the accuracy of these statements. For cells at least, telomerase *was* a fountain of youth. But, of course, what the reporters wanted us to speculate about was whether the immortality gene could be used to slow the aging process in the whole human body. If we could find a way of extending the telomeres in all the cells in the body, what would

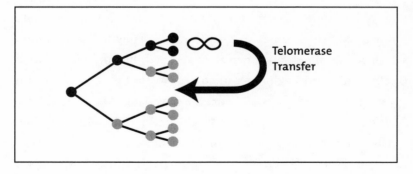

Figure 11. Telomerase Transfer. The transfer of an active gene for the catalytic component of human telomerase into human cells can make them immortal. Can this technology be used to treat aging in living human beings?

happen? We simply didn't know the answer to that most important question. And it was outrageous and irresponsible to speculate about such a sensational idea.

So the reporters went with what they had. And it was enough for great newspaper headlines. The *Oakland Tribune* headline read, "Researchers 'Immortalize' Human Cells." The *New York Times* article was titled "Cell's Life Stretched in Lab." The Palo Alto paper read, "Menlo Firm Finds 'Fountain of Youth' for Cells; Its Stock Soars."

The *Los Angeles Times* headline read, "Scientists Give Cell Apparent Immortality," and the article went on to say, "Breaking the biological barrier once thought out of reach, scientists have for the first time endowed healthy human cells growing in a dish with a quality that alchemists, explorers and mystics have vainly sought for ages: immortality."

But after all the hyperbole, I had to ask myself: "Are we any closer to understanding and intervening in actual human aging?"

Chapter Six

The Philosopher's Stone

But the lengthening of the thread of life itself, and the postponement for a time of that death which gradually steals on by natural dissolution and the decay of age, is a subject which no physician has handled in proportion to its dignity.
—FRANCIS BACON
The Advancement of Learning

Before there were chemists there were alchemists. Just as astrology was a nonscientific and superstitious precursor to modern astronomy, so alchemy served as the forerunner to chemistry. But alchemists were unique in that their crucibles burned with a passion for a singular and focused objective. They labored day and night to find a substance they called the "philosopher's stone." This imaginary red substance was believed to contain the power to transform common substances like lead into precious and incorruptible gold. The alchemist observed that while the hand that wore a gold wedding band was mortal and old age destroys beauty, the gold ring itself was immutable and incorruptible. It did not rust or tarnish over time. As a result, gold became elevated to a symbol of immortality.

From the Hebrews' menorah to the cup of the sacrament of Eucharist, gold has long been a symbol of the wish for immortality. Mankind hoped to live forever like this simple metal. So the alchemist sought the philosopher's stone, the substance that could transform other substances into gold, both for the wealth it could provide them and also for the ability to transmute a corruptible human body into an incorruptible and eternal one.

Early scientists, including such luminaries as Isaac Newton, tinkered with alchemy. Newton received from the hand of the famous chemist Robert Boyle a red candidate for the philosopher's stone, but, by the ironies of fate, the substance most probably contained mercuric oxide, and instead of opening the gateway to immortal life, it could well have been the cause of mercury poisoning and his early death. Recent studies have found traces of mercury in remnants of Newton's hair.

In another, perhaps more straightforward approach, some early gerontologists simply took the immortal end product, gold itself, and tried to assimilate its properties. This led to the concept of a gold elixir *(aurum potabile)*. It was Roger Bacon, well aware of the profundity of the problem of human aging and death, who endeavored to extract from gold the secrets of its incorruptibility. His most notable application of the alchemists' technology was a "gold liquor" for Pope Nicholas IV, a vain attempt at achieving pontifical immortality.

Over time, the science of chemistry grew in stature and scientists distanced themselves from the search for the philosopher's stone. Indeed, chemists were as embarrassed by any association with alchemy as astronomers are today with astrology. As science matured, it recognized that what distinguished it from other areas of inquiry was that it was committed to the rigorous pursuit of data that can be measured, to observations that can be reproduced by others—in other words, to a strict and critical adherence to observable facts. And so science, and what later came to be called scientific medicine, eschewed the faith healer, the snake oil salesman, and the fortune-teller. And anyone claiming to be able to reverse the aging process was *a priori* outside the legitimate realm of science.

So you can easily imagine the storm clouds of criticism that arose following the newspaper headlines announcing that we had found the immortality gene. The use of the word "immortality" to describe the property of cells being able to replicate without limit is unfortunate. First used by August Weismann in the 1800s, it never was intended to imply that the cells in question could not die. It was meant only, in a scientific sense, to describe cells that *need* not die, cells that, when maintained under proper conditions, with adequate nutrition and so on, had the potential to proliferate without any genetically limited life span.

But the use of "immortal" in this scientific context was too deeply entrenched for us to change it, and we found ourselves stuck with the word. So telomerase would from this point on be known as the "immortality enzyme." The scientific community may have understood the context, but the general news media waxed eloquent. In many an interview, I was asked, "Will telomerase one day make you and me immortal?"—the type of question that inevitably made me squirm in my chair.

The easy answer would have been to quickly deny any use for telomerase in extending the human life span. Indeed, many gerontologists, ever quick to distance themselves from any possible association with what others might call pseudoscience, went so far as to say that science would *never* change the way we age. But I kept trying to tell the truth as I saw it, not just to give an answer that lent an appearance of respectability. Over and over again, in those first few weeks after isolating the telomerase gene, I said that we really didn't know how much of human aging was due to cellular aging and therefore might be treatable in principle with telomerase therapy. It could turn out to be anywhere between 5 percent and 90 percent of what we call human aging. But even it were only 5 percent, and we could change that 5 percent, wouldn't we have done something wonderful?

The real answer required that the same experiment be performed on people, as opposed to cells in a dish—that is, the telomerization of the human body or, at a minimum, the telomerization of specific human tissues. The telomerization of cells in a dish was a very simple experiment once we'd found the telomerase gene, but the introduction of genes into actual human beings to cure disease, called "gene therapy," has been notoriously problematic. Only a small percent of cells in the dish actually absorb and regulate the gene in a reliable manner. This high level of inefficiency is acceptable for cells grown in the dish because there are ways of clearing away the undesired cells and propagating only the modified cells so as to produce millions of cells modified in a specified manner. But it doesn't work that way in the human body.

Let us suppose that we wanted to test the effect of telomerase on aging skin. And let us assume that we could paste the telomerase DNA onto the skin and then electrically zap it into our skin cells. The reality is that only a small percent of our cells would take up the DNA and use it in the way desired. And, of course, such profound "tinkering" in an actual human being would raise all kinds of concerns about safety—one of the most obvious of which would be to make sure that, in telomerizing a tissue, we didn't damage the DNA and raise the risk of cancer.

So it was clear that although we had "immortalized" the human cell, we didn't have the technology to go the next step. That left us in the twilight zone of showing that telomerase could halt the aging of human cells, but without our being able to demonstrate its utility in real human aging. Nevertheless, this didn't slow down profiteers who, seizing on the fad, raced to market with new telomere skin creams to tap into the dream of eternal youth.

Performing an actual telomerase therapy is still theoretically possible. One of the more promising ideas has been to insert the telomerase gene into a virus. Viruses are like microscopic syringes. In their evolution, they have developed the ability to inject their own DNA or RNA into cells, thereby subverting the cell and forcing it to turn its molecular factories into making copies of the virus instead of normal healthy proteins. So one of the key ideas in gene therapy research has been to utilize this remarkable feat of nature's engineering by exchanging a desired gene for a viral one, then injecting it with its new therapeutic gene package.

In addition to the practical problem of actually getting the telomerase gene into the cells of the human body as I mentioned before, there is the worrisome issue of whether telomerase therapy might in reality be an excellent means of causing cancer.

My own hunch is that telomerase therapy, assuming it could be properly administered, need not lead to cancer. How can this be true? To find an answer, we need to go back to the runaway-car story.

As Robert Weinberg and others have shown us, cancer cells have several different things broken in their genetic program. They have a broken accelerator (or oncogene), stuck to the floorboards, that tells the cell to proliferate as quickly as possible. Like a car with the accelerator stuck at full throttle, speeding through stoplights, without the ability to properly regulate its speed in traffic, such a cell will wreak havoc on the normal functioning of a human tissue. The out-of-control cells will accumulate and form a growing mass that we call a tumor.

But in addition to a broken accelerator, tumor cells frequently have broken brakes as well, mutations in the tumor suppressor genes. Such cells cannot arrest their growth as they normally would when there is serious DNA damage in a cell. Like a car with a stuck accelerator and a broken brake, cells with both mutations are close to becoming a real problem.

But a runaway car will eventually run out of gas. And the multiple-step nature of cancer requires that cells proliferate many times before they can experience multiple mutations. Such extensive proliferation will generally lead to telomere shortening and senescence before it can become a dangerous and malignant cell. That is probably why, as we age, our skin is covered with so many moles and pigmented spots. They were a pigmented cell screaming out of control, but they never immortalized. They ran out of gas, and the cells grew old. But what if the cell had a broken accelerator, a broken brake, and an infinite fuel supply? That would be a particularly deadly combination.

Weinberg eventually proved this to be the case in a landmark experiment. His lab took normal cells and then, one step at a time, made a series of defined changes in the DNA. He mutated a gene to make an oncogene, he broke the tumor suppressor gene, and he added telomerase. The cell was now cancer! It was the first time in history that mankind could demonstrate in such a dramatic way that we really were beginning to understand this enigmatic disease. We could now make a cell cancerous.

But with simple telomerase therapy, we wouldn't be creating this deadly combination. A car with an infinite fuel supply would not necessarily be a runaway car. Indeed, the ideal telomerase therapy would only extend telomere length and then cease, much as the refilled gas tank in our automobile only takes us another few hundred miles. Cells that had their telomere length restored without permanent immortalization could in fact treat age-related diseases and at the same time actually reduce the risk of cancer.

But this is—and was—more theory than fact. I already knew in the mid-1990s that we would not likely, anytime soon, have a means of delivering telomerase efficiently into the human body. So where did we go from here?

I asked myself this question countless times without a good answer. But one sunny morning as I walked along San Francisco Bay, still trying to find my way out of the labyrinth, I phrased the question differently. What, I wondered, if we could find a population of cells that could be rejuvenated by telomerase outside the body and then returned to it? My mind went back to the branching diagram of human cellular development and the dichotomy of the germ line and soma. The more primitive the cell we telomerized, the more different kinds of young cells we could make for the aging body.

My mind flashed back to a day in my old med school pathology lab, where, once a week, we students enjoyed a brief respite from our books and the opportunity to actually work with diseased tissues removed from the operating rooms of the hospital. One afternoon, our professor walked into the lab, stainless-steel bucket in hand, but instead of pulling out the typical lung tumor or cirrhotic liver, he removed a small—maybe two inches in diameter—nodule of tissue.

"This," he announced, holding up the tissue the way a preacher would hold a Bible, "is a benign tumor called a teratoma."

Teratomas are most often diagnosed when a woman comes to her doctor with pain in her abdomen. On studying an abdominal X-ray, the doctor asks the patient whether she has swallowed her dentures. If not, she has a teratoma. This unusual structure is a mishmash of teeth and hair, blood and bones, brain and skin all wrapped up tightly in a ball. It is not a pregnancy; it is instead a bizarre experiment of nature gone awry. It never comes close to being anything like an organized developing human being.

Our professor, that afternoon, picked up his scalpel and sliced the teratoma in half. There, inside the mass, under the fluorescent lights of the lab, glistened two pearly teeth, an incisor and a premolar.

Amazing, I remember thinking. What aberration of cell biology could have led to such an amazing phenomenon?

If the appearance of the conglomeration of teeth and hair had a disturbing appearance, years later the memory set loose a speculative chain in my mind. What if we could somehow re-create the miracle of the teratoma in the laboratory? And, if we could do that, what if we could control it in such a way as to make the specific cells and tissues—heart, pancreas, lung, liver, and kidney—that people who were sick so desperately needed?

The critical questions to answer were: (1) what kind of cell had gone wild in order to make the teratoma? and (2) how could we isolate and grow that cell in the laboratory dish to recapitulate the process in the laboratory?

I had an idea. The germination of new life from the ovary looked to me as if it had sprung from a very primitive sort of stem cell. A stem cell, as its name implies, is a cell that can "branch out" like the stems of a tree and form more than one cell type. At minimum, to be a stem cell, as opposed to all of the cell types that inhabit most of the tissues in our bodies, a cell needs to be able to divide into two cells, one of which will be another stem cell similar to the original and one of which will change, or "differentiate," into another cell type.

The cells on the surface of our skin are like this. They are called keratinocytes. The specks that slough off our skin, dandruff, for example, are all many hundreds of keratinocytes. There is a stem cell in the skin that divides, making another stem cell and a keratinocyte in order to continually repopulate the surface of our skin. This type of stem cell is called "unipotent."

But there are also "pluripotent" stem cells—that is, cells able to differentiate into multiple cell types. The best-characterized pluripotent stem cells reside inside our larger bones and are therefore often referred to as bone marrow stem cells. They are capable of branching during cell division to make red blood cells, the various white blood cells such as B and T lymphocytes, the garbage collector cells known as macrophages (literally, "big eaters"), and other cells as well. So the bone marrow stem cells are more "powerful" in the sense that they can become more different kinds of cells than just a skin stem cell. But it appears that pluripotent stem cells from an adult human being are always somewhat limited in what they can make.

I wondered whether we ever might be able to find such a thing as a "totipotent" stem cell—one that was able to make any cell in the human body. Was such a cell even conceivable? It would be less like a stem, more like the trunk of the tree of cellular life. It would be the mother of all stem cells. It would be the raw material from which people are made.

Of course, such a cell had never been isolated from the human being, but if it were, it could revolutionize medicine. It could allow us to make any kind of cell or tissue in the laboratory dish. And this, in turn, would open the door to much new and important research, and of course, potential therapies to replace cells or tissues lost or dysfunctional as a result of disease.

In the years following my first exposure to the teratoma in medical school, I'd tracked down a lone researcher named Leroy Stevens who, years before, had found the phenomenon as interesting as I did. Stevens had worked at the mouse research center called the Jackson Laboratory in Bar Harbor, Maine. In 1953, he was working under a grant from a tobacco company that suspected that it wasn't tobacco but cigarette paper that caused cancer. To test this, he exposed mice to both the burning paper and the tobacco and looked for the rates of cancer in the two groups. One day he noticed a bizarre tumor in the scrotum of one of the mice. When he dissected it he found an assortment of teeth and hair—a teratoma! This particular strain of mice known as 129 seemed to be predisposed to this strange type of cancer. Stevens spent the rest of his life pursuing this biology in his strains of 129 mice and dropped the tobacco research.

Stevens bred the 129 mouse to select for a high propensity to form this type of abnormal growth. He then turned his attention to the important question of determining what type of cell could make teratomas. By looking earlier and earlier in the genesis of this marvelous phenomenon, he traced its origin to the primitive germ-line cells that were in the twelve-day-old embryonic testis or ovary. They were on their way to becoming sperm and egg cells but were still in a relatively undifferentiated state. They could still become any cell in the body.

Then, through careful observation, he noted that these cells were very similar to the undifferentiated cells in the week-old embryo, the blastocyst stage when it is just a little ball of unformed cells. When Stevens transplanted these most primitive embryonic cells into the scrotum, he again got teratomas! In other words, he had succeeded in finding the cells that generated the teratoma. They were primordial germ-line cells, either in the first stages of the embryo or slightly later, just as the cells were becoming sperm or egg. He called this class of cells "pluripotent embryonic stem cells."

In the years following Stevens's studies, a scientist at the University of California, San Francisco, named Gail Martin showed that the cells of the very primitive mouse embryo could be cultured in the laboratory and maintained as a continuous culture of totipotent stem cells. Cell cultures that can be continuously and reliably cultured are called cell lines, to distinguish them from cells that can be grown for only a brief period of time. At about the same time Gail Martin reported her work in 1981, Martin

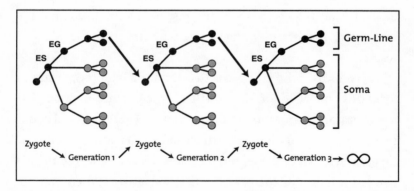

Figure 12. The Discovery of the Germ-Line Cells Called Embryonic Stem (ES) Cells and Embryonic Germ (EG) Cells in Mice. ES cells come from the pre-implantation embryo, EG cells from the later-developing gonad.

Evans and Matthew Kaufman at the University of Cardiff in Wales also isolated mouse embryonic stem ES cells.

Then, in 1991, Brigid Hogan's and Peter Donovan's labs, again working independently, managed to isolate embryonic stem cell lines from the 12-day-old testes and ovaries of mice (the cells first studied by Stevens). Their careful study of culture conditions showed that signals could be identified that caused the cells to multiply in the dish while staying in the undifferentiated germ-line state. Peter and Brigid came to name these cells *embryonic germ* (EG) cells, to distinguish them from the more primitive ES cells that came from the embryo when it was simply a ball of unformed cells.

These cells, either ES or EG cells, were just the kind of cells I wanted to try to isolate from human beings. If allowed to grow as a three-dimensional mass of cells, they could perform the magic of forming the very same teratomas, the balls of hair and teeth, the "grab bag" of cells and tissues I saw back in medical school, the cells so many people needed so desperately.

As I walked along the Bay that sunny day, these technologies mixed together in my mind. What if we could telomerize the human totipotent stem cell, the types of cells that Stevens had called "pluripotent embryonic stem cells"? Then couldn't we make young cells of any kind, at any point on the developmental tree? This, it struck me, would give us the greatest range of possibilities for making some cells that could be telomerized and put back into the human body to treat aging.

But then I stopped in my tracks.

Weren't these cells naturally immortal? Weren't they *zoë*, not *bios*? Didn't they belong to the immortal germ line? Were they not akin to the immortal cyclings of the sun worshiped in the ancient world, the source of the immortal renewal of life?

The primordial stem cell on the left in Figure 12, because it is totipotent, has yet to make the early "decision" on whether it will be a mortal somatic cell or an immortal germ-line cell. As a result, one would predict that the human primordial stem cell, if we could find it, would stay young in the dish, that it would be immortal and telomerase positive, and that it would begin to age only after it began to differentiate into a particular kind of cell. So just as babies are born young, any cell made from a primordial stem cell would be born young.

If this worked, we wouldn't even need telomerase!

I felt that this was an important idea. These cells were the green essence of human life, forever branching to make you and me.

This was the path I had been searching for.

But did I really want to get into *that* area?

There is an area of gerontological research that is largely taboo, one tainted by a long history of chicanery. It is known as "rejuvenation science." Early in the twentieth century, after Alexis Carrel circulated the wrong view that all of the cells in the body are naturally immortal, gerontologists went back to the drawing board to think up new explanations of how and why we age. If the somatic cell did not age intrinsically, then it seemed logical to conclude that aging was caused by something outside of the cell, like the loss of the proper milieu for the cell to thrive. In direct support of this idea, Carrel reported that blood taken from old people was less able to support the growth of cells in the laboratory as compared to that from young people. This naturally led to the idea that perhaps there was the loss of some "youthful factor" in the blood during aging that causes cells to no longer function properly. Perhaps this essential substance was a hormone, a chemical messenger that travels in the blood, and its loss caused our cells and bodies to malfunction.

Even Carrel could not deny Weismann's point that the reproductive cells escape this aging, allowing us to make babies that are born young. Fellow transplantation surgeon Serge Voronoff concluded that there was a circulating substance, a hormone made by the organs that house the reproductive cells, that was lost with age. He argued that because the hormone was concentrated in these organs, the reproductive cells were protected from the age-related loss.

There were likely other reasons Voronoff came to this conclusion. Some thirty years earlier, Charles Edouard Brown-Séquard had experimented in hormones produced by the testes. Brown-Séquard had established himself as a prominent scientist with his contributions to our understanding of the hormones manufactured by the adrenal gland. On June 1, 1889, in a scientific lecture on hormones at the Société de Biologie in Paris, he shocked his audience by announcing that he had rejuvenated himself with extracts of testes—that is to say, injections of testicles ground up in a salt solution. In his own words:

> Before beginning my formal lecture, I must tell you of my
> latest experiment. I am 72 years old. My natural vigor has
> declined in the last 10 years. I have injected myself . . .
> [and] today I was able to visit (rendre visite) young
> Madame Brown-Séquard . . . [the injections have] taken
> 30 years off my age.

The new elixir, this *liquide testiculaire*, was the talk of the Parisian salons. It seemed credible because of the stature of Brown-Séquard, and perhaps also because of the immutable hope of conquering death. And perhaps because of the weight of the antiquity of the theory behind it. Going all the way back to the ancient world, extracts of testes were used to impart life. In the Far East, the ancient Chinese recommended *coitus reservatus*—that is, frequent sex with young women, but sex interrupted prior to ejaculation. The concept here was simply to keep the immortal essence within the body.

The Viennese physician Eugen Steinach advocated a surgical version of *coitus reservatus*—namely, tying off the tiny tubes through which the sperm flow, a technique that today we call a vasectomy. This, he argued, would force the body to keep the gonads full and active in rejuvenating the body. In 1920, Steinach published the work *Rejuvenation Through the Experimental Revitalization of the Aging Puberty Gland*. The power of the gonads to rejuvenate was gaining popularity. The Steinach procedure was performed on hundreds of patients, including such luminaries as Sigmund Freud. And the breadth of interest threatened even the Church, to the extent that a Swedish Church leader threatened Steinach with everlasting damnation as a result of his efforts to play God.

So it is understandable that Voronoff, who had originally experimented in transplanting ovaries in an attempt to treat infertility in women, happened upon the idea of using these same techniques in rejuvenation. Since the new scientific approach to medicine was to carefully test ideas to demonstrate that they actually worked, Voronoff went to the laboratory bench. In 1917, he surgically removed the testicles of a young ram and transplanted them into the testicles of an old animal. To his astonishment, he saw, or at least he believed he saw, a real difference. The old animal, which had long before lost an interest in the opposite sex, was again bleating the sweet refrains of romance. This was not, of course, clear evidence that aging itself had been reversed. It was, if anything, likely the result of an elevation of the male sex hormone testosterone, so named because it is made in the testicles. But Voronoff was impressed enough that he decided to discuss his results publicly. On October 18, 1919, at the Twenty-eighth French Surgical Conference, he announced, "I have found a remedy for old age. I have already rejuvenated a number of animals." So Brown-Séquard, Steinach, Voronoff, and their followers mobilized to market the medicine to restore youth through the testicles. The group of scientists, nicknamed the "Erector Set," proclaimed the maxim: "You are only as old as your glands."

Spurred on by this apparent success, Voronoff took the next step and launched one of the more controversial experiments in the history of science. Because human testicles were hard to come by, he instead removed the testicles of a monkey, sliced them in wedges like the slices of an orange, and implanted them into the testicles of a man. At the Thirty-first French Surgical Conference, he presented his first "rejuvenated" patient, Sir Arthur Liardet. And the patients and testimonials began to accumulate.

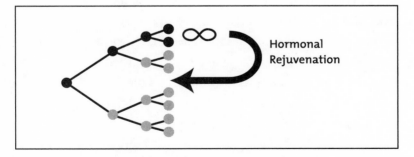

Figure 13. Voronoff's Testicular Therapy. Believing Carrel's data suggesting that somatic cells do not age intrinsically, Voronoff tried to reverse aging by transplanting young monkey testicles into an old man in the belief that the rejuvenating hormones would make the whole body young.

Understandably, Voronoff's testicular therapy ignited a fire-storm of controversy. His claims regarding rejuvenation itself would have caused a stir, but the details as to how the rejuvenation was achieved could hardly be kept secret. His growing monkey business raised the eyebrows of many investors, for Voronoff rapidly became the focus of hope for the aging people of his generation. The Paris paper *L'Humanité* stated that the talk of the town in 1919 was rejuvenation more than politics.

The transplantation of animal testicles and ovaries never fully caught on, especially when it became clear that they would be rejected by the patient's immune system as foreign tissue. But the search for male and female hormones extracted from ovaries or testicles led to isolation and synthesis of testosterone by Adolf Butenandt's and Edward Doisy's isolation of estrogen, both in 1935. When it was confirmed that estrogen levels did indeed fall with age, women were given the purified hormone to determine whether menopause, and perhaps other manifestations of female aging, could be put on hold. There have been many claims over the years for hormone replacement therapy, hopes that it could be used to slow the onset of skin aging, bone loss, heart disease, and Alzheimer's disease. While the results continue to be controversial, it has become clear that the loss of these hormones is not the actual cause of aging. Even with hormone supplementation, people continue to age and die.

Only a few brave souls performed research on aging in the period 1930–1960. A notable example was Viennese physician Paul Niehans. Niehans began his career observing the sunset of Voronoff's. According to Niehans, Voronoff came to him in the final months of his life and confessed the "horrible discovery" that he had infected many of his patients with syphilis. Apparently, the monkeys were infected and the disease had been passed on during the transplantation process. Indeed, Voronoff died a defeated man. But Niehans was to learn from the mistakes of the rejuvenators. He decided to use cells from a more distantly related animal, the fetal sheep. And rather than transplanting the gonads, he proposed transplanting young cells of many kinds. The theory was that if you had diseased kidneys, he would take a fetal lamb, extract its kidneys, mince them up, and inject them in your posterior. His Clinique la Prairie in Vevey, Switzerland, was to become the leading rejuvenation center of the world. It was to treat such leading personalities as Pope Pius XII. Niehans's career skyrocketed in 1954 when he appeared to have cured the Pope with cell therapy. Indeed, the results were so impressive that the Pope's personal physician, Dr. Galeazzi-Lizzi, later went into a private practice in fetal cell therapy himself. As he said: "When an organ is sick, as from old age, its cells seem to be tired of reproducing themselves; the cadence varies, the form changes. What else can one do against this degeneration except bring fresh cells to the tired organ?"

But Niehans had to have known that sheep cells would be immediately rejected by the patient's immune system. Perhaps in Voronoff's time this was not clear, but it was in Niehans's time. As a result, the rejuvenators tarnished the reputation of gerontologists for decades. During those years, aging research became a black art, a great way to destroy one's reputation in the scientific community.

I was mindful of all this history as I walked by the Bay that morning, and I knew full well that an attempt to isolate the naturally immortal human primordial stem cell from fetal testicles would stir up enormous controversy and disapproval, both inside and outside the scientific community. But the logic was solid. If we could isolate these cells and grow them at will in the tissue culture dish, we could potentially make young cells and tissues of any kind to replace those failing from old age.

By sheer luck, I had an appointment that same afternoon with a scientist named Dr. Roger Pedersen at the University of California, San Francisco, medical school. I had arranged the meeting ostensibly to talk about collaborating to make embryonic stem cells (ES cells) from a special strain of mice that made telomere studies much easier to perform. Because ES cells come from an embryo at one of the earliest points in development—before a pregnancy has developed and before the embryo has begun the process of differentiation to make the cells of the soma—it is called a preimplantation embryo. In view of the prevailing controversies, let me explain exactly what that means.

In the process of in vitro fertilization (IVF), scientists perform the earliest stages of conception in the laboratory dish—hence the somewhat pejorative term "test-tube babies." The first step in the process is to obtain egg cells from the surface of a

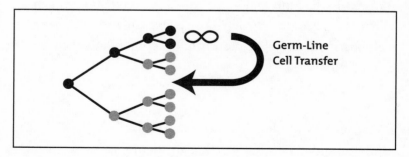

Figure 14. Cell Therapy for the Treatment of Aging. Cells from the immortal germ line, if cultured in the laboratory, have the potential to make many kinds of young cells to replace those worn out with age.

woman's ovary. This is usually accomplished by giving the woman shots of hormones that cause her ovaries to release between ten and twenty egg cells in one month instead of the normal one egg cell. These are removed and placed in a small laboratory dish with a liquid cell culture broth to feed them. Then sperm cells are added to the dish. The sperm cells swim around until they bump into an egg cell. With a little luck, they stick and then begin to burrow into the outer leathery shell of the egg. The first sperm cell to enter the inner part of the egg cell triggers a chemical reaction that then prevents a second sperm cell from entering. Fertilization is complete, and the resulting cell is called a *zygote*.

Over the next few days, the new cell will begin the process of cell division. It will become two cells, then four, eight, and so on. At this point of development, the embryo is extremely small. It can comfortably fit on the point, let alone the head, of a pin. It is normally traveling down the fallopian tube at this stage, on its way to finding a home in the uterus. Nearly two weeks after

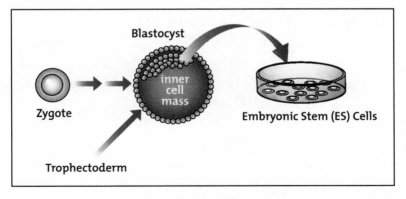

Figure 15. The Source of Human ES Cells. ES cells are derived from the inner cell mass of a blastocyst embryo prior to the implantation of the embryo in the uterus and prior to the beginning of differentiation.

fertilization, the preimplantation embryo "hatches" out of the tough shell surrounding it, and if it can attach to the uterus and burrow into this warm, nurturing environment, it will "take root" and a pregnancy will result. Until implantation occurs, there will never be a positive pregnancy test, and about 40 percent to 60 percent of such embryos formed naturally in the body never attach and simply die. Typically, another 10 percent or so are lost soon afterward.

The embryo at this very early stage is called a *blastocyst;* it is really a microscopic ball of totipotent, but not yet committed, stem cells. Surrounding this mass of totipotent stem cells is a sac of cells, called the *trophectoderm cells,* that are on the "lookout" for the uterus. If they find the uterus and attach, development will begin. The trophectoderm will become the placenta and the totipotent stem cells, called the *inner cell mass,* will become the entire animal. If the blastocyst does not find the uterus, then the ball of cells will not begin to develop and there will not be a pregnancy.

The cells of the inner cell mass are all still blank cells that have not begun any step in forming differentiated cells. They can, for instance, still form identical twins. Therefore, ethicists who study embryology are very comfortable with the idea of preimplantation embryos being either implanted or discarded in the way sperm is discarded if a pregnancy is not desired.

Roger Pedersen was not only an expert in making mouse ES cells, he also directed a human IVF clinic at the UCSF Medical School. I wondered, as I drove to our meeting, if I might be able to convince him to collaborate in trying to isolate the human ES cell from spare human preimplantation embryos.

How would he view the idea that those that were going to be thrown away anyway could be used to try to grow human embryonic stem cells? This might be easier than isolating them from fetal testicles.

I walked onto the beautiful campus overlooking the Golden Gate Bridge. This was the campus where, years earlier, Gail Martin had isolated some of the first mouse ES cell lines. I knocked on Roger's door, and he greeted me warmly. We sat down and talked about yet another mouse project. I explained about my interest in telomeres and aging, and why I would like to get ES cells from the inner cell masses of this special strain of mice. Roger understood the project. Elizabeth Blackburn, Carol Greider's old mentor, was now a faculty member at the university, so he had heard all about telomeres.

The ease of our conversation, and Roger's disarming manner, led me to chance broaching the taboo subject.

"Roger, have you ever thought of making human ES cells?" I asked quite innocently.

He seemed almost insulted by the question, and as if a fog were suddenly rolling in across the Golden Gate, I could feel a chill in the room. Roger started to answer, but then he seemed to catch himself. "Yes, I have considered that possibility."

And that was that. He was already halfway out of his chair when he finished the sentence, and he led me off to the mouse room in what appeared to be a conscious effort to change the venue, and with it, the subject.

What I failed to appreciate at the time was the level of sensitivity that still lingered in the field of human embryo research following the controversy surrounding Robert Edwards and Patrick Steptoe, who produced the first test-tube baby in the 1970s. Just a year earlier, in 1994, President Clinton had been pressured into passing legislation to ban all federal funding for anything that could be characterized as human embryo research. As a result, human ES cells were the subject of hushed conversations in the hallways during scientific meetings. The mutual understanding was that secrecy was the best policy. I, in contrast, always operated under the assumption that if a technology was good and just, then it should be proclaimed from the rooftops, not whispered about in dark corridors.

Nevertheless, Roger and I proceeded with our mouse project.

By scrambling around and persuading, I found a means of getting early human fetal testes and tried to grow the human EG cell in a dish, duplicating the conditions that had worked with mice. Despite my diligent efforts, the cells would not proliferate. They just sat there in their dish, looking back at me.

I thought I knew what was wrong. My fetuses were sixteen weeks old, far too old for this purpose. I needed five-week-old fetuses. But where could I get those? Women do not abort fetuses that early, when they are just learning they are pregnant.

Time passed, my frustration grew, and then, one day, Roger Pedersen called me at Geron.

"Mike, do you remember the time when you proposed collaborating to make human ES cells?"

"Sure I do," I answered. "Why do you ask? What's up?"

"Well, I've thought about this very carefully, and I could use some private funding for such a project. Would you still be interested?"

Private funding meant Geron. And that was the day the project to isolate the human ES cell was launched.

Shortly thereafter, I learned that a scientist at the University of Wisconsin in Madison named James (Jamie) Thomson was about to publish a paper in the scientific journal *Proceedings of the National Academy of Sciences* on the isolation of the ES cell from a species of monkey. On hearing this, I realized that having access to the monkey cells could prove to be the key step on the path to isolating the human cells. In addition, it was possible that the U.S. Patent Office would grant a patent on the cells and extend it to all primates, including humans. This would give us proprietary rights on the human version even before the human cells were isolated.

Within a week, I flew to Madison. My first impression of Jamie was of an unusually quiet and dedicated scientist in his late forties, as unassuming as his office. I immediately liked him. I explained our work with telomeres to him and why I was interested in ES cells. He, it turned out, had used the same technique with monkeys that Martin and Evans had employed back in 1981 to grow the first mouse ES cells. He was eager to pursue his work with human tissue, and my arrival couldn't have been better timed in that regard. He was well aware of the ban on federal funding enacted in the previous year. In order to continue, Jamie not only had to obtain nonpublic funding, but he would need a whole new separate laboratory facility, with everything purchased with private money, right down to the test tubes.

I camped out in a Madison hotel, laboring to forge a collaboration with the university and Jamie. Providing funding for Jamie wasn't the problem. A small company like Geron could turn around such documents on a dime. The conflict was that my company expected to get exclusive rights to any patent that might come from the work. But WARF, the Wisconsin Alumni Research Foundation, which controls all exploitations and applications of their scientific work at the University of Wisconsin, refused. Not only did they refuse, but they did so on a nonnegotiable, take-it-or-leave-it basis.

As with all successful business transactions, we compromised. I now had access to preimplantation embryos through Roger's IVF clinic and, through Jamie's skills, to rhesus monkey cells. Jamie mailed me the cells, and we had the privilege of being among the first to study ES cells from a species close to human.

The cells were magic. Just as I had hoped, they "bloomed" in the laboratory dish into all of the cells of the body. In normal development the embryonic cells branch into three primordial lineages: ectoderm—from which such cells as hair and the photosensitive cells called rods and cones in the retina arise, mesoderm—from which cells like muscle and bone originate, and endoderm—from which cells like the iodine-hungry cells of the thyroid or the ciliated cells that slide mucus over our respiratory system arise. These cells "blossomed" in the same way in the dish, although not in an organized fashion that would lead to a developing animal.

Nothing like this had ever been seen before in the history of science. The monkey ES cells became everything, and if allowed to grow in three dimensions, became actual complex tissues like intestine, teeth, hair, eyes. They made what one of our scientists wittily described as "a dish of rhesus pieces!" I saw in these new cells a power not unlike that of atomic fission. They could be harnessed in hundreds of ways to help sick people.

One day I walked into the tissue culture lab and Andrea Bodnar (who had led our project to immortalize human cells with telomerase) met me at the door excitedly.

"Mike, the cells are beating!"

I walked to Hayflick's microscope. There in the corner of the dish of cells was a small area, perhaps the size of a pea, rhythmically beating on its own, contracting, pulling the cells along with it in the rhythm of life. In the view of a cell biologist, this is no big deal. The contraction of myocardial cells is easily explained. It is not that much different from the motions of any other cultured cell. Still, seeing this small bit of heart tissue took me back to that night I sat in my lab at home, wired up to the oscilloscope, wishing I could help my father.

"Take a video of it while we have it," I told her.

That afternoon I had a meeting with a chairman of a department at UCSF. I wanted to show him the video as a way of communicating how powerful these cells were. In a small conference room with the seasoned surgeon, I began with a scholarly presentation on the cells, then I clicked on the video. He had no doubt witnessed many hundreds of beating hearts in the thoracic cavity of patients. Still, when he saw the video clip of the beating heart we had made in vitro, he leaned forward in his chair, wide-eyed.

"Oh, my God!" he whispered reverently.

Could he see what I did—what they could do for medicine? If we could get the human cells growing, we could make cells in the human body—young cells that could be the basis for hundreds of new medical therapies. But even more, could he appreciate their grandeur—the immortal lineage of cells that have streamed down the millennia? They were once in monkeys in the Pleistocene, the Miocene; they were in the primitive mammals of the age of dinosaurs. I wanted so badly for someone to see what I did in these cells. They were life itself, the immortal thread that connected the generations. And they were the *fons et origo* of all the cells in the body. I felt that if people could understand what these cells were, they would talk of nothing else. People would literally stand at the street corners and ask one another, "Have you heard of the immortal cell? Isn't that just amazing!"

I called Jamie and told him about the reaction I had at UCSF.

"You know, Mike, I don't think this is a good idea, showing that video."

"Why not?" I asked quite innocently. "I think he got it after seeing them."

"Well, you know, I have visions of people protesting outside of your facility, with T-shirts saying, 'Stop the Killing of Little Beating Hearts.' I think it could get out of hand."

Maybe Jamie was right. But I so desperately wanted people to see what I saw—a power that could heal a broken heart, that could give children a cure for diabetes, that could regenerate cells and tissues that cannot normally heal themselves, like the nerves in the spinal cord and brain.

Despite Jamie's reservations, I used the video and put together a presentation to a group of venture capitalists. My hope was to raise a hundred million dollars to fund this new branch of research. I invited Roger Pedersen to come and talk about our efforts to get the human cells.

"Gentlemen," he began, "I believe something magical is about to happen." And then he took the investors through the work he was doing on our behalf, ending with a personal story. He told them of a dear little girl who lived next door and was a friend of his daughter. This innocent child fell ill one day and her concerned parents took her to the doctor, where their worst nightmare became a reality. Their little girl had leukemia. Roger then leaned forward with a penetrating gaze, fixed the eyes of the investors, and with the force of a judge delivering a verdict said, "Gentlemen, I would have done *anything* to save the life of that girl. *Anything.*"

The depth of his emotion resonated in the room, and I could

feel exactly what he meant—the mix of rage and determination. He didn't literally mean *anything,* of course. He wouldn't have murdered a fellow living and breathing human being to save the girl, but he certainly would have risked his career, family finances, reputation. He would have risked his own life. *Let them whisper in the hallways,* he seemed to be saying. *I'm going to do the research that would have saved that precious little girl.*

I, too, gave a presentation. I showed the investors our new data indicating that the monkey ES cells were telomerase positive and that they appeared to be naturally immortal. This, I explained, was the first isolation of the immortal cell from a primate. And then I showed them the beating heart.

I had given many hundreds of corporate presentations and could easily take the temperature of an audience. I looked around the room. I saw glazed eyes. The investors didn't get it.

"Mike, stem cells have been around for a long time," one of them said, "and no one has built a successful business with them."

"Yes, but these aren't just any old stem cells. . . ."

"It probably isn't even possible to grow a human cell," Tom Maniatis, a well-known scientist, advised. "If it were possible, it would have been done by now."

I couldn't dent them. After all, these were hardheaded businessmen looking for profits. And Roger's and my passion didn't have the smell of money.

So I went back to the lab. We would just have to grow the human embryonic stem cells, I concluded, then try again with the investors. Unfortunately, my own attempts at culturing a human EG cell line from testicles were failing. Roger and I talked it over and agreed that the problem was that I needed five- to eight-week-old human abortuses. The testicles and ovaries I was using were too old. The cells were well on their way to becoming sperm and egg cells. They had lost the magical totipotency.

Roger told me of a scientist at the Johns Hopkins University School of Medicine in Baltimore, Dr. John Gearhart, who had arranged a collaboration with a Maryland abortion clinic whereby such early tissue samples could be obtained and used in research. John had an interest in a medical problem called trisomy 21, a DNA abnormality that leads to mongolism and mental retardation. He believed that by isolating cells from the developing egg cells, he could potentially gain an understanding of why the disorder occurred.

Once again, I was on a plane. But I had one stop to make first. My father had a tradition of bringing his mother roses on his birthday, one rose for every year of his life. It was his way of saying thank you for bringing him into the world. It was a Friday afternoon, and my birthday was that Sunday. I changed my itinerary at the last minute and flew into the airport in South Bend, Indiana, rented a car, and drove to my mother's home in southwestern Michigan. I surprised my mother at the door with forty-three red roses.

My mother and I spent a gentle weekend together reminiscing. I kissed her goodbye at the door, knowing that there were not many more roses I would bring her because of her age. The fog lay low over the farm fields as I drove back to South Bend.

I had to work harder and faster—the clock was still ticking.

John Gearhart couldn't have been more different from Jamie Thomson. He was as outspoken and outgoing as they come. He was all for our collaboration, but it took a full eighteen months for me to negotiate the contract with John's university before he could join in our collaboration.

While I struggled to get the contract in place with Johns Hopkins Medical School, I continued to experiment with the monkey ES cells. I was an officer and board member of Geron, but I couldn't get more than two scientists to work with me on the ES project. These were the years, of course, when we were closing in on cloning and testing the telomerase gene, and nothing else seemed as important to the company as seeing the telomerase projects through to completion. Many of the company's financial backers saw the ES cell project as a distraction.

When, in 1996, we took the company public, it was on a platform of telomerase inhibition as a cancer therapy, telomerase therapy for age-related disease, and research into human primordial stem cells (as I generically referred to human ES and EG cells). The impetus for the initial public offering, to be sure, came from our venture-capital investors, who were eager to cash in. They and our Wall Street underwriters sent me and two of my colleagues on a private-jet whirlwind tour of the United States and Europe to present the company to interested parties—a surreal experience that didn't quite sink in until the morning came when, back at last in my North Beach apartment, I saw on the financial channel the letters "GERN ... GERN ... GERN ..." flash across on the ticker at the bottom of the television screen. "GERN" was Geron, and we were now on Nasdaq. Suddenly the company's success in making products took on a new level of tension.

The following year we had a double success, announcing first the successful testing of telomerase and then the news that Gearhart at Johns Hopkins had isolated the first human ES cells (more accurately, EG cells). The rewards of success in science, especially success on this scale, are gratifying indeed, and I had every reason to enjoy the limelight.

But one fateful day, sitting in a staff meeting, I realized how disillusioned I had become about the progress I was making. The topic under discussion certainly didn't help. As I recall, it had to do with the allocation of reserved parking places in the lot outside our building. I was now forty-four years old. I had so much to accomplish and so little time. We were nowhere near finding the path out of the labyrinth. If I were truly to succeed in deciphering human aging, which is what I had set out to do years earlier, then I needed to run faster, not lean back and debate parking places. I remember thinking that if I were on my own and had access to a good medical library, I could get more accomplished than I currently was at Geron. And once I thought the thought, it was but a small step to hearing myself say it aloud to my wife, Karen.

"I think I ought to leave Geron," I announced to her one evening soon afterward. "If I can get more done that way, I should leave."

Impulsive as I may make it sound here, it was no easy decision. Geron, in the beginning, had been my means to accelerate scientific discovery into aging, and as a fruit of this research we likely had the patent on the immortalizing gene telomerase. We had such a wonderful collaboration to isolate the natural immortal cell, the human ES cell. And I loved the company we had built, and valued the friendships with the scientists among my colleagues. But somehow I knew that I was losing my struggle to instill in the company my vision to focus on gerontology.

Nothing was going to stand in my way, and I saw anything that slowed me down as an adversary.

Karen agreed with me completely.

Forty-eight hours later, I was gone.

Shortly before I left Geron, I visited John Gearhart at Johns Hopkins—partly to celebrate his triumph in culturing human

ES cells, but also to discuss where the research would go from here. We spent a good portion of the day sitting in his office discussing the collision course we were on with the religious right. John had a deep appreciation of the political turmoil embryo research had generated in the previous decades; he knew many of the scientists that had been "burned at the stake" for their research in the field. I, on the other hand, because of my years of study in fundamentalist Christianity, could see the motivations of people on both sides of the issue. Even so, John and I saw eye-to-eye on where this was all going. This research was going to unleash one of the fiercest battles between religion and science in recent history. We both knew it, but neither of us relished the prospect of the fight, a battle that he, Jamie Thomson, and Roger Pedersen were going to have to face without me.

Then—I'll never forget it—John proudly walked me into his tissue culture lab, where, reaching into the incubator, he produced a small dish of the cells. As I brought them into sharp focus peering through his microscope, I saw what I could not isolate before—an island of flat, glistening cells. *This is human life,* I thought to myself. *Not a human life, but human life. All of these cells are poised and prepared to participate in the creation of an individual human life.*

For the first time in history, we were looking at human *zoë. Zoë,* not *bios.*

The dream of the ancients.

The immortal cell.

I was deeply moved. This, I knew once again, was where my own path lay. Walking from the building, I admired the antique and august portraits that line the hallways of Johns Hopkins. Among them was one of William Osler, the first physician-in-chief of the budding medical school in 1888. Osler, one of the greatest physicians in history, wrote the book *The Principles and Practice of Medicine*, the standard textbook of medicine for the next forty years. He traveled to Harvard University in 1904 to give a lecture titled "Science and Immortality." The subject of his talk was the impact of August Weismann's description of the immortal germ-line cells.

His words echoed in my mind as I walked from the building:

The patiently worked out story of the morphological continuity of the germ plasm is one of the fairy-tales of science. You who listen to me to-day feel organized units in a generation with clear-cut features of its own, a chosen section of the finely woven fringe of life built on the coral reef of past generations—and, perhaps, if any, you, citizens of no mean city, have a right to feel of some importance. The revelations of modern embryology are a terrible blow to this pride of descent. The individual is nothing more than the transient off-shoot of a germ plasm, which has an unbroken continuity from generation to generation, from age to age. This marvelous embryonic substance is eternally young, eternally productive, eternally forming new individuals to grow up and to perish, while it remains in the progeny always youthful, always increasing, always the same. Thousands upon thousands of generations which have arisen in the course of ages were its products, but it lives on in the youngest generations with the power of giving origin to coming millions. The individual organism is transient, but its embryonic substance, which produces the mortal tissues, preserves itself imperishable, everlasting, and constant.

Chapter Seven

Human Therapeutic Cloning: A Maelstrom of Controversy

The use of embryos to clone is wrong. We should not, as a society, grow life to destroy it. And that's exactly what's taking place.
—PRESIDENT GEORGE W. BUSH

Mad scientists are still mad scientists no matter how white their lab coats are and how many bioethicists they have to justify their actions.
—REPRESENTATIVE CHRIS SMITH
USA Today

"B okanovsky's Process," *repeated the Director, and the students underlined the words in their little notebooks.*

"One egg, one embryo, one adult—normality. But a bokanovskified egg will bud, will proliferate, will divide. From eight to ninety-six buds, and every bud will grow

into a perfectly formed embryo, and every embryo into a
full-sized adult. Making ninety-six human beings grow
where only one grew before. Progress."

These words in the opening pages of Aldous Huxley's novel *Brave New World* shocked the sensibilities of many people. Huxley described a cold and unfriendly world where people were cloned to make specified, defined, and predictable classes of workers. Was this what science and technology were coming to, he seemed to be asking, a mechanization of life where human beings are manufactured for purely utilitarian ends?

Bokanovsky's process was pure fiction when *Brave New World* was published in 1932. Some sixty years later, however, a young cell biologist named Keith H.M.S. Campbell, responding to an ad, took a position at a research center called the Roslin Institute, situated in the hills outside of Edinburgh, Scotland. Keith had been intrigued by the concept of cloning ever since 1984, when, at the University of Sussex, he had sat in on a lecture by Karl Illmansee. Illmansee, an accomplished and respected biologist, claimed to have succeeded in cloning mice, but his own behavior and the inability of others to reproduce his results eventually led many to doubt the validity of his work. In 1984, Davor Solter and James McGrath stated in *Science*: "The cloning of mammals by simple nuclear transfer is biologically impossible."

But Keith believed otherwise. Defying the evidence to the contrary, he loudly and often asserted to his lab mates that he would someday clone a mammal.

The concept of cloning is quite simple. It begins with an egg cell, a cell that is in many respects like any other in the body. It contains DNA, but unlike other cells, it normally jettisons half of its DNA in order to "marry up" with a similarly halved male sperm cell. As a result, a new cell emerges that has a normal amount of DNA but is a meld between the two parents. In the cloning process, the egg cell's DNA is simply removed by means of a microscopic glass needle about the diameter of a hair. Then the DNA of a somatic cell is transferred into the egg cell so that all of the genetic information comes from a single parent. The result is a clone, an identical twin, though born at a different time from the animal it came from.

Cloning first made headlines in 1978 in the wake of the birth of the first test-tube baby, Louise Brown. The same year, David Rorvik authored a best-selling novel titled *In His Image: The Cloning of a Man*. Rorvik, previously a science writer for *Time*, claimed to be telling the inside story of the cloning of a sixty-seven-year-old millionaire. Scientists in the field of embryology knew the novel was likely a hoax, but Rorvik refused to provide any proof as to the authenticity of his story or to admit that it was fiction.

Nonetheless, images of renegade scientists huddled in secret underground laboratories, funded by eccentric millionaires, were deeply troubling. Rumors spread of new efforts in Singapore, North Africa, and South America. Should they be believed? Well, such a story was certainly more credible than accounts of bodies of aliens frozen in an airplane hangar in Roswell.

As a result, an international controversy erupted.

Jim Watson stepped into the debate largely in an effort to defend the cloning of DNA. There were many reckless calls for legislation that would ban the whole field of recombinant DNA research. When asked whether Rorvik's story was credible, he answered, "Cloning—particularly of something so complex as a mammal—cannot be done with an adult cell, not to mention a cell taken from an eccentric sixty-seven-year-old." Later he summarized: "If either of my young sons wanted to become a scientist, I would suggest he stay away from research in cloning humans. There's no future in it."

Watson was accurately reflecting the scientific consensus of his day. Keith Campbell's later conviction that the DNA in any of the body's cells could be used by the egg cell to make a clone was a faith not held by the scientific world at large. But he reasoned that, since virtually every cell in the body has the identical copy of DNA, a body cell, if transplanted back into the environment of an egg cell, could be "reprogrammed"—that is, changed in such a way that it acted as an embryonic cell. If this worked, then conceivably any cell in the body—say, a skin cell—could be transferred back into an egg cell to reprogram it into an embryo, and then the embryo placed into a uterus to make a pregnancy and eventually a live animal that would be identical to the animal from which the cell was taken. In other words, it would be a clone. But when Keith said that he was going to create a clone of a sheep from a somatic cell, even his coworkers at the Roslin Institute were certain he'd gone bonkers.

Ian Wilmut, the head of the Roslin cloning effort, had much more practical concerns. He was collaborating with a nearby company, Pharmaceutical Proteins, Ltd., known as "PPL." PPL's interest was in making genetically engineered animals. They hoped to modify them in such a way that they would secrete pharmaceuticals into their milk. This would make their milk worth millions of dollars. If they succeeded, and Roslin's scientists were inventors on the patents, there could be a tidy financial reward.

Up to this point, all of the methods of genetically modifying animals had proved quite inefficient. Ian's hope was to clone from sheep ES cells. The logic of the experiment was clear. Ian had learned from Neal First's lab at the University of Wisconsin in Madison that cattle could be cloned from ES cells, even those that had been grown extensively in a laboratory. Because they could be grown in a laboratory, they could be genetically engineered in the dish by zapping DNA into them, just as we had done with the telomerase gene. These engineered cells could then be used in a cloning experiment. Ian thought this might work because mouse ES cells could be genetically modified quite easily.

Furthermore, since ES cells were not differentiated, the scientists at Roslin wouldn't be asking the egg cell to reverse time's arrow. The prevailing wisdom among scientists all over the world was that undifferentiated germ-line cells become soma and die. There were almost no exceptions, certainly none in such advanced animals as the mammals. Time's arrow always points to aging and death once the commitment is made to branch off into somatic cells. It's as if the "decision" to make muscle, brain, and bone carries with it a commitment to ingest a suicide pill.

Ian Wilmut therefore resisted Keith's forays into using differentiated cells in nuclear transfer. But Keith was stubborn. So he hit on a ruse. He designed the experiment with a control. As I mentioned in discussing my days in Sam's lab, controls are often performed, but all too often late in the course of an experiment. So Keith argued that he should do a control right at the start. He would include some cells that came from the ES cells but that had started to differentiate. These would be what is called a "negative control." That is, they should *not* work. But as I said, his was a clever ruse. Keith thought they actually *would* work, and this was his way of proving it.

Keith and his team at Roslin transferred the ES cells and the differentiated cells derived from ES cells into enucleated ewe oocytes. The first good news came in February 1995. A pregnancy check was positive: Fetuses had come from the differentiated cells!

In August 1995, Megan and Morag were born, two sheep cloned from primitive cells that had differentiated from ES cells. The scientific world took note of this, as it was the first case of mammals cloned from anything other than an undifferentiated cell. Time's arrow had now been reversed in a mammal, at least a little—from a barely differentiated cell back to the beginning of life. But the bombshell came in the next round of experiments.

The following year, Keith wanted to go for the real thing. Because Roslin was collaborating with PPL, which was studying the expression of milk from breast cells, Roslin just happened to have some breast tissue cells growing in the lab from a six-year-old female sheep. With the excuse that these cells would make an even better "negative control," Keith included these fully differentiated adult-derived cells.

The team again went to work performing the nuclear transfers. They produced 277 reconstructed embryos from the adult breast tissue cells. Twenty-nine of them successfully developed into blastocysts, which were then transferred into thirteen ewes, only one of which led to a live birth. But the point is, one of the breast cells *did* become a warm, fuzzy, bleating lamb! In other words, a mammary cell that had once been programmed to express the proteins for milk now reexpressed the genes of the embryo, and the end result, on July 5, 1996, was the miraculous birth of the lamb Dolly.

Keith was right and all of the scientific community was wrong. An animal could in fact be cloned from *any* living cell in the body. Time's arrow, at least the arrow of cell differentiation, could be reversed!

Keith was away on vacation that July 5 when he got the

phone call. The lamb had been born. After some discussion, she was named Dolly, after Dolly Parton—as a tribute to the striking fact that she was cloned from breast cells.

Keith wrote up the details of the experiment in a four-page paper and sent if off to the scientific journal *Nature*. The paper was accepted, and publication was scheduled for February 27, 1997. As it happens, articles that make it to such prestigious journals as *Science* and *Nature* are considered by virtue of that fact to be newsworthy. Therefore, about a week before they are published, they are shared with the news media in the form of a "tip sheet" explaining the paper and its implications. This allows the members of the press to do interviews and write intelligent and, it is to be hoped, well-thought-out articles for publication on the day of the announcement. But in the case of Dolly, the first cloned mammal, there was a strong suspicion the embargo might break—that is, that someone might publish the story ahead of the release date. Sure enough, on Sunday night, February 23, Robin McKie of the *Observer* found that the story had leaked, and he immediately released the article to the public. The media around the world then rushed to the presses in a panic to report the story on Monday.

The world's media converged on Roslin. Keith Campbell slipped into the shadows, and Ian Wilmut, who was a media natural, became known around the globe as the leader of the Dolly team.

The announcement of the cloning of a mammal from a somatic cell immediately reawakened the controversy from the 1970s. The public's perception, fanned by sensation-seeking media, was that this was a real-life playing out of *The Boys from Brazil*. Who could say what the next step would be? The cloning of Tutankhamen? Or Adolf Hitler?

From the moment I started in the race to clone the telomerase gene, I'd never taken the time for exercise or even, for long stretches, moments of true relaxation. Now, in January 1998, I was determined to take a break. I would spend some months in deep study of the scientific literature and organizing my next foray into aging research. Meanwhile, in my first day as a free agent after leaving Geron, I drove to the San Francisco Bay Club to work out in the weight room. I stopped at a sporting goods store across the street and began piling up tennis shoes, T-shirts, and shorts for the next hundred sessions in the gym. "You certainly are stocking up," the store's owner remarked. "This is the first day of my retirement," I answered with a laugh.

An hour later, I was in the weight room. With some twenty pieces of equipment and the room practically to myself, I determined to start at one corner and work my way to the other side, one station to the next. I adjusted the weights and began lifting. Turning my head, I watched myself in the mirrors filling the wall. There I was, a perfect reflection of myself, a clone. My mind drifted off to Dolly and the miracle of nuclear transfer. What excited me about cloning was not that you could now theoretically clone your favorite hunting dog or racehorse. As a cell biologist, the revolutionary concept to me was that it was possible to turn back the clock on a somatic cell, taking it back to an embryonic state. It was so unexpected. As far as we knew, nature simply never did that. At least not in mammals like humans. So why should it work? It was like the sand of an hourglass falling upward, or a wineglass falling off a table onto the floor and then reversing its motion with the pieces coming together and landing on the table, intact all over again without a drop spilled.

This was as close to a miracle as we ever see in science.

Cloning shouldn't work. But it did!

What, I wondered, if we could take an old human cell for a ride in this time machine? Not to clone a human being but to take it back to an embryonic state and, let's say, make ES cells? We could theoretically reverse the differentiation of a human cell. From these embryonic cells that would now be totipotent again, we could fashion cells of virtually any type in the human body and they would be identical to the original patient's cells.

It has been estimated that 3,000 people die every day for lack of transplantable tissue. Although many people offer their organs for transplantation, shortages are acute. Moreover, transplantation of tissue from one person to another does not solve one basic problem: The human body rejects tissues that are not its own. Essentially, doctors are forced to "trick" the body by finding an available organ that is a close match to your tissue type, and then, using a battery of drugs, to anesthetize your immune system so that it fails to "notice" that the transplanted tissue is not you.

The use of cloning in medicine would mean that we could make any kind of tissue identical to the patient by taking one of the patient's own cells back in the time machine to make it embryonic again, then differentiating the ES cells into whatever the patient needed—heart cells for a failing heart, liver cells for patients with cirrhosis, pancreatic cells for diabetes, blood cells for cancer patients, and so on. In addition, by starting with a cell easily grown in the laboratory, like a skin fibroblast, it would be possible to introduce precise genetic modifications and then, through the use of this time machine, make ES cells and then any genetically modified cell the body needed.

Needless to say, this was a powerful concept.

But then this line of reasoning—logical enough—joined with another. What if an aged somatic cell could not only be reprogrammed by the egg cell to become an embryonic cell, what if, just what if . . .

I dropped the weights.

We are, after all, returning a cell to the immortal germ line. What if an old cell with short telomeres were returned to the germ line—that is, transferred to the egg cell by cloning. Perhaps the telomerase in the germ line would repair the damage of age and reset the telomere length, making the old cell truly young again in regard to cell life span! This would be an *incredible* result, potentially a true reversal of the aging of a cell. The "reversal of aging" demonstrated in the first cloned animals was merely a demonstration of the reversal of differentiation—that is, a skin cell (differentiated and mortal) was transformed into an embryo. And, of course, that embryo, when implanted in the uterus of a mother, then became a cloned animal that looked like

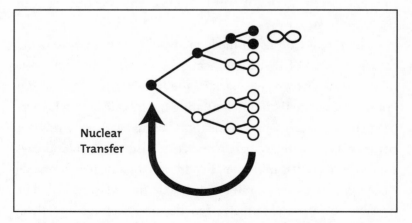

Figure 16. Is Cloning a Cellular Time Machine? The transfer of a somatic cell into an egg cell can reprogram the somatic cell back to an embryonic state. But does it reverse the aging process in the somatic cell as well?

the original. But what if the cloned embryo was like the original animal's embryo? Are cloned animals born as old as the original cell, or does the clock of aging get reversed as well?

If it reversed aging, this was a *monstrously* big idea!

I won't go into the details of the aging process, except to say that nuclear transfer could be expected to reverse other aspects of aging as well. For instance, the batteries of the cell, small entities called mitochondria, are damaged in aging, but the cloning process would introduce new egg-cell mitochondria, just as happens in normal reproduction.

This, I saw, was the experiment of a lifetime. I dropped the weights, showered, and never again wore any of the T-shirts. I saw my path in front of me again, leading directly out of the old one. I had a lot of work to do, and far too little time.

In an effort to forge some kind of collaboration with the Scottish laboratory that had cloned Dolly, I scheduled a meeting with Ian Wilmut. Karen and I flew first to London's Heathrow Airport and then flew in a second, smaller plane north to Scotland, to the city of Edinburgh.

The following day, we took a cab to the Roslin Institute and waited in a cafeteria, where we were to meet Ian Wilmut. When he arrived, we walked over to a nearby table and I started to give him the lowdown on my background in telomere biology and about my efforts to isolate the human ES cells. This led quite naturally to a brief discussion about therapeutic cloning. The second I had finished my "icebreaking diatribe," Ian launched into a similar speech about Roslin's programs in animal science, and then, of course, about Dolly. We had both probably given such little talks thousands of times, and we continue to do so. It is an opportunity to share information, but more important, to learn something about other people: Are they intelligent? Are they easy to work with? And so on. Ian seemed to me a rather mainstream scientist caught up in a maelstrom far beyond what he expected—by which I mean an affable and no-nonsense sort of fellow. He was also very, very stressed. He made repeated references to all the lectures and television appearances he had done in the past year, a lifestyle he was not accustomed to. As he walked us around the lab, though, introducing us to other scientists, he seemed to calm down, as though for him the world of science was more familiar, more relaxing.

I asked him the question burning in my mind.

"Have you considered measuring Dolly's telomeres?"

Ian was aware of our recent article in *Science* proving that telomere shortening caused the aging of human somatic cells. "Well, we've thought of it, and hope to collaborate with Howard Cooke here in Edinburgh to do the work." How strange, I thought to myself, were the twists and turns of the labyrinth! For it was Howard Cooke's data on telomere shortening that had convinced me, once upon a time, that telomeres were the clock of cellular aging. It was thanks to him that I'd jumped from the medical school train six years earlier.

Ian and I agreed to continue discussions about setting up a new company to focus on the human applications of cloning. We would meet next in Australia, where he was scheduled to deliver a paper at a cloning conference at Monash University in Melbourne.

On Thursday, April 16, 1998, I sat next to my wife in the back row of the auditorium at Monash. Ian presented first, recapping what everyone already knew, that Roslin had cloned sheep from cells soon after they had differentiated from ES cells and also from adult breast tissue. The next speaker was Jim Robl from the University of Massachusetts at Amherst.

Jim was a tall, rugged-looking man—half scientist, half cowboy. His talk was mostly about cow cloning, which was where cloning research was focused in the United States. The cow was a more significant animal than the sheep from a commercial point of view, and Roslin would have used cows, too, if it hadn't been for their high cost. Maintaining cattle was simply too expensive a proposition for the Roslin Institute. Sheep are cheap.

Jim Robl then described his efforts—and the efforts of one of his graduate students—to genetically modify cells before they were used in cloning. This procedure generally requires that cells divide many times before cloning, and so it is limited by the natural life span of the cells. But the graduate student had used genetically modified cow cells in a cloning experiment after they had expended almost their entire life span.

I leaned forward in my chair.

The cells, originating from a 40-day-old fetal cow, would go 30 doublings. The genetically modified cells were now about 29 doublings old. Nevertheless, the student had used these cells in a cloning experiment and gotten embryos. He'd then implanted the embryos in a mother cow and 40 days later gone back to retrieve the fetal cows.

Jim then told us that the fetuses looked altogether normal, and that they had led him inexorably to the big question: Could cloning reset the clock on cellular aging? To see, he had put the cloned fetal cells in culture again and measured how many doublings the cells would go through now. He showed us all in Melbourne a bar graph of the results. The cloned cells had gone about 30 doublings, the number expected of normal young cells!

I leaned over to my wife, sitting next to me in the lecture hall, and whispered.

"Oh my God!"

I was awake. It was a life-changing moment, for I knew full well the implications of what Robl was telling us that day. Cells like fibroblasts from large mammals (like humans and cows) age as inevitably as death and taxes. Unless it was an outright fraud (and I had no reason to think that was the case), this data suggested that cloning actually *had* reset the clock of cellular aging. Suggested, but not proved, and, after all, it had been done with cow cells, not human—but still, my instinct was that we should all feel very lucky about what the egg cell was offering us.

As soon as the last talk in the session wound to an end, I headed out of the auditorium in the direction of Jim Robl. He was easy to spot, as he stood head and shoulders above the crowd. He was chatting with someone about lunch plans when I buttonholed him.

"Jim, I enjoyed your talk. I was interested in the data you showed on the restoration of cell life span by nuclear transfer. Have you measured telomeres? This would be the definitive test."

"No," answered Jim curtly. "You should call Abel Ponce de Leon about that."

Ponce de Leon? Was he putting me on?

"Who?" I asked.

"Abel Ponce de Leon," he repeated. "Abel is with our company, Advanced Cell Technology. You should call him."

I knew that Jim had been one of the pioneers in cloning, inventing many of the techniques that Keith Campbell and Ian Wilmut were now using routinely in the lab. But who was Abel Ponce de Leon?

When I returned home, I looked up Abel on the Internet.

He was on the faculty of the University of Massachusetts at Amherst, and was working on ES cells from chickens. This made sense, because ACT was owned by Avian Farms, a chicken genetics company in Waterville, Maine. Avian's business was cloning animals for agricultural purposes.

Then I hit on an idea. I had a good excuse to knock on the door of Avian Farms. In a spin-off of our ES cell efforts at Geron, Karen and I had formed a company called Origen Therapeutics to use chicken ES cells as a way of genetically modifying poultry. The goal was to make the chicken that could lay the golden egg—eggs full of pharmaceutical proteins like antibodies, blood proteins, and hormones. I called Rob Saglio, president of Avian Farms, and expressed an interest in talking "chicken." He agreed that we should talk, and he set up a meeting at ACT in Worcester, Massachusetts.

A few weeks later I was sitting in the offices of ACT. I spoke first, talking about chicken biotechnology. Then, during a break, Rob Saglio pointed to a mild-mannered, dark-haired young man who'd been sitting at the table.

"Jose," he said, "why don't you tell Mike about your work?"

Jose turned out to be Jose Cibelli. He was a recent graduate from Jim Robl's lab, now employed full-time at ACT. But far more important, as I realized from the slides he began to show us with a projector, he was the same young scientist whose data Jim had just presented in Australia!

Again I sat forward in my chair.

It seemed Jose could read my emotions, because he skipped over the slides on why ES cells and cloning could be important and went directly to a picture of human ES cells. Then he dropped a bombshell.

"These are ES cells I made using my own cells and a bovine oocyte."

The somatic cells he'd worked with, in other words, were his own! He had taken a glass slide, he confessed, and scraped the inside of his cheek and isolated the live cells under a microscope. He then used nuclear transfer, combining them with the egg cells of a cow.

Did he have any concept how scandalous this was? I wondered.

I explained to him how just a couple of months earlier a scientist named Richard Seed had stood up at a cloning conference and declared to the world that he was going to clone himself. This somewhat brash gesture had led to near panic in Washington, with rash legislation drafted overnight to ban all nuclear transfer using human cells. I explained that I had testified at a House hearing in February 1998 and worked with Senator Kennedy's staff, and that when Senator Lott had brought the bill to a vote on the floor of the Senate, Kennedy had argued eloquently on behalf of patients with Parkinson's disease, diabetes, heart disease, arthritis, and cancer. The bill had gone down in flames. At that time, no senator was prepared to go on record as one who had blocked important medical advances. But Jose's experiment, if news of it got out, could revive the hysteria.

"Jose, in those hearings the Congress asked me whether anyone had actually performed nuclear transfer with human cells," I said. "I told them no. The last thing this controversy needs right now is to hear that it has been done, performed in secret, and that the egg cells were not human, they were cow!"

Jose just shrugged his shoulders.

I was very concerned, though—not about the embryo, but about the way it had all been done. If the story ever leaked to the media and was told by someone less than sympathetic to our cause, the result could be disastrous. I could see *60 Minutes* knocking on ACT's door with the camera running. "Dr. Cibelli, is it true that you cloned yourself?" I could imagine the newspaper headlines: "CRAZED SCIENTIST CREATES HALF-HUMAN, HALF-COW CLONED EMBRYO!" We had enough problems in the biotechnology industry raising capital for risky long-term projects, while dealing with an encyclopedia of regulations in getting our products approved by the FDA. The last thing we needed was a scandal that, if handled the wrong way, could measure 9.0 on the Richter scale and put all kinds of burdensome regulations on our work.

Still, I was convinced that ACT, not Roslin, might be the best base from which to launch a human therapeutic cloning project. For the next several weeks I negotiated with Rob about a collaboration working on chickens, then slowly introduced the concept of a merger of ACT with Origen. The management at Origen was enthusiastic about a collaboration on chickens with Avian Farms but didn't share my vision about the human therapeutic cloning project. Again I found myself at odds with a company over a matter of vision. The management of Origen wanted to make the chicken that laid the golden egg; I wanted to use the egg to rejuvenate human cells.

But I simply couldn't let anything stand in my way. I was out to determine whether cloning really could be used as a cellular time machine, and if it could, I would find a way to translate it into a means of rebuilding the old human body with young cells.

One day in June 1998, I called Rob Saglio and broached the question directly. Would he consider my coming to ACT as its CEO? We negotiated quietly for a few weeks, and then one night in September, while Karen and I vacationed at the Villa San Michele in Florence, Italy, I faxed Rob my signature on an agreement. I felt the way I had the day I cashed the first investment check to Geron; I was now committed to making ACT and the human therapeutic cloning project a reality. Three months later, Karen and I took our last ride across the Golden Gate Bridge in San Francisco, leaving a peaceful life in California for a future in the eye of a storm.

In Greek mythology, the Minotaur is a half-human, half-bull creature that guards a maze called the labyrinth. The labyrinth, with its branching underground caves, is a symbol of the residence of the dead. Once a person dies and enters this realm, there is no escaping; no one finds their way back out to the light of day. To make matters worse, this fearsome creature, the Minotaur, is there to torture its citizens. According to the legend, one brave warrior named Theseus successfully entered the labyrinth and slayed the Minotaur. He escaped with the aid of his lover, Ariadne, who gave him a ball of string to unravel on the way in, thereby providing him a means of finding his way out.

On my first day at work in October 1998, I began to deal with the puzzle of what to do about the Minotaur embryo experiments. Three things came to mind. First, this data had been submitted in a patent application and was technically in the public domain (people in the media, however, don't make a habit of reading patent applications). Second, ever since the publication of our paper at Geron on telomerase rewinding the clock of cell aging, and particularly since Woody Wright's "cellular fountain of youth" comment, I had become a regular phone call on the Rolodexes of print and broadcast journalists out to make sense of the story for the public. "So, Mike," they'd ask in the course of conversation, "won't you tell us about what you're working on now?"

The third thing was that, right at that moment, we were not doing the work in question. After the Dolly uproar, President

Clinton had asked for a voluntary moratorium on human cloning, making no distinction between reproductive and therapeutic cloning. We didn't have to respect that moratorium request—it didn't have the force of law—but as an American, I felt that we should respect it. But did the President really intend to restrict the medical side of cloning, or just the cloning of people?

No one knew.

At the same time, I believed passionately in the potential medical uses of nuclear transfer, and if that meant using a non-human egg cell to reprogram a human cell, where was the harm? I thought it needed to be pursued, and pursued with all haste. But I didn't want to violate Clinton's moratorium.

From the beginning I was determined to do this right. I could live with criticism, but not criticism based on any misrepresentation or clandestine research on our part. Consequently, I decided that the best solution would be to be completely transparent about what we had achieved and what our plans were. In addition, I needed to form an Ethics Advisory Board to give us an arm's-length ethical review on our projects. If and when we started the human cloning work again, it would be with flawless and rigorous ethical review.

So when *48 Hours* called us hoping to do a program on the future of medicine, I agreed. I knew that they were preparing for "sweeps week" in December and were looking for a sensational program that would make their ratings competitive with other networks', but still I hoped for balanced coverage. As we filmed in ACT's labs, I tried my best to communicate my vision of how the egg cell could take a patient's cell back in time. When asked point-blank about the use of a cow egg cell, I explained how the use of animal egg cells would reduce medical costs, and to allay fears about half-human/half-cow embryos, I explained that we removed the DNA from the cow's egg cell and that using the empty egg cell as a receptacle would make us no more part-cow than drinking cow's milk would make us part-cow.

When they were done, I lectured them like a grade-school teacher.

"The lives of thousands of sick people lie in your hands now," I told them. "You need to take what you've just filmed with the utmost seriousness."

I hoped for the best. But later, as I walked back into the lab, I felt something under my shoe in what was normally a spotless and virtually sterile environment.

"What's this?" I asked a technician as I reached down and picked up what appeared to be a piece of hard-boiled egg.

"Didn't they tell you?" the technician answered. "I thought it was odd at the time, but they asked whether anyone had a hard-boiled egg. Actually, we did. Then they asked for some liquid nitrogen. They dipped the egg in the liquid nitrogen, ultrafreezing it, and then filmed it as they dropped it on the floor."

"They dropped it on the floor?"

"That's right. And it shattered into a million pieces."

"Did they explain why they would do such a crazy thing?"

"Something about how the use of human eggs in cloning could shatter the fabric of society."

I'd been around long enough to understand that the media can skillfully spin an interview a number of different ways. They can paint you as either a hero or a huckster. But there were human lives at stake this time.

Two weeks before the *48 Hours* program was scheduled to air, I got a call from Nicholas Wade of *The New York Times*. "What do you think about Jamie Thomson's and John Gearhart's papers?" he asked.

"What papers?"

"They are publishing the isolation of human embryonic stem cells in next week's *Science* and *Proceedings of the National Academy of Sciences.*"

This was bound to come sooner or later. John Gearhart had actually announced his cells in 1997, before I even left Geron. But I couldn't think of worse timing for these papers. If Jamie's and John's papers were published the week before the release of our news, it would appear that we were simply trying to say "Me too!" Who would possibly know or believe that the *48 Hours* program was my effort at managing a highly controversial piece of research and had been planned many weeks earlier?

But there was no way of stopping the train at this point. Jamie's and John's papers were announced on November 5, 1998. "Scientists Cultivate Cells at Root of Human Life," read the *Times* headline. The article began: "Pushing the frontiers of biology closer to the central mystery of life, scientists have for the first time picked out and cultivated the primordial human cells from which an entire individual is created." Then, a week later, after I'd given Nick Wade an interview, the *Times* wrote a less flattering article on our cross-species cloning work. As it turned out, *48 Hours* was mild compared to the *Times*. I would have fully expected Nick to produce a low-key treatment about the potential advantages of cloning in personalizing the human ES cell. Instead, his article opened with the words "Venturing deep into uncharted realms of ethics and medicine . . ." and went on: "The technique would involve creating an embryo of uncertain moral status, and one that crosses the barrier between humans and other species. . . . [T]he concept of half-human creatures arouses deep-seated anxiety, as is evident from the unfriendly powers ascribed to werewolves, centaurs, mermaids, Minotaurs and other characters of myth and folklore."

Two nights later, I sat at my computer, plowing through my e-mail, when I stumbled on a letter from President Clinton to the National Bioethics Advisory Commission. "This week's report of the creation of an embryonic stem cell that is part human and part cow raises the most serious of ethical, medical, and legal concerns. I am deeply troubled by this news of experiments involving the mingling of human and non-human species. I am therefore requesting that the National Bioethics Advisory Commission consider the implications of such research at your meeting next week, and to report back to me as soon as possible."

The Commission was set to meet on the following Tuesday in Miami. I had to be there.

I walked into our bedroom.

"What's going on?" Karen asked, reading my face like a book.

"Oh, just that the President is calling for an investigation into our work. It's nothing really. Let's go to sleep. Oh, by the way, I'm going to have to fly down to Miami on Monday."

A couple of days later, I walked into the large meeting room of a Miami hotel and sat as inconspicuously as I could in the back. It didn't work. One of the members of the presidential commission was none other than Carol Greider, my previous collaborator on telomeres. She cornered me in the back of the room.

"Mike, you sure know how to stir up controversy."

"Well, we are just trying to advance nuclear transfer in medicine," I answered.

I testified to the Commission, laying out my vision of what good could come of such basic research, but the panel was clearly way ahead of me on these issues. One member said with a smile, "Mike, just go ahead and do the work. Clinton's moratorium should be interpreted to refer to cloning babies, not cells."

On November 20, the Commission responded to Clinton: "An attempt to develop a child from these fused cells should not be permitted." But in regard to the use of nonhuman egg cells in human therapeutic cloning: "Combining human cells with nonhuman eggs might possibly lead someday to methods to overcome transplant rejections without the need to create human embryos, or to subject women to invasive, risky medical procedures to obtain human eggs."

Although I had generated many more waves than I intended to, the outcome, I thought, was a success. We had completely defused what had been a ticking time bomb. We had been completely transparent in what we had done, and now we had been advised, informally, to proceed. And that is what we did.

The first real work on human nuclear transfer began in January 1999. At that time, I thought we were competition-free—until, that is, on May 4, 1999, I read the press release from Geron announcing the purchase of Roslin's spin-off company, Roslin Bio-Med, which had the licenses to exploit the Institute's patents commercially. One of the announced goals of the acquisition? To clone human ES cells. I was now in competition with Ian Wilmut and my old company.

And then, less than two weeks later, on May 26, 1999, Geron and Roslin announced that Dolly's telomeres were prematurely short. Ian Wilmut and the scientists at PPL wrote a paper published in *Nature* claiming that Dolly had been born old! Because Geron held the patents to telomerase, Tom Okarma, Geron's new CEO, was able to proclaim that Geron—and Geron alone—could fix the shortened telomeres with telomerase. Geron's press release stated, "Geron and the Roslin Institute will collaborate to combine telomerase activation with nuclear transfer to enable the efficient production of cloned transgenic animals with normal, full-length telomeres." It was clear what this implied: ACT could clone only geriatric animals, since we did not have the telomerase patent.

Ian granted an interview to the BBC:

IAN WILMUT: There are mechanisms that bring our lives to an end. And what we have been able to do by looking inside the cells is look at one of those mechanisms. You may remember that she [Dolly] was produced from a cell taken from a six-year-old ewe. She herself is now nearly three years old, so the question we have asked is, is this mechanism set for a three-year-old ewe, or a nine-year-old?

REPORTER: And what's the answer?

IAN WILMUT: The answer is for nine years. That particular clock has not been reset during the process of cloning.

REPORTER: So she is older than she is?

IAN WILMUT: That particular mechanism is, yes. Now, we have no idea of the consequences. It is quite probable, we believe, that she herself will have a normal life and there will be nothing different. But if we were to clone from a clone and then do the same thing again and again, at some point, and we don't know when, it would begin to have a deleterious effect on the clones. It is obvious that we need to do a lot more research to understand this.

REPORTER: Well, of course, you realize that people who are against the whole business and wherever it will lead will seize on this and say here you are, you are playing God, you are getting into worlds you don't understand, this is just a glimpse of, you know, a Frankenstein future. That's what they would say, wouldn't they?

IAN WILMUT: Well, we would share some of those concerns because we would only wish to use the technology as it is proved to be safe, as we understand it more, so we would share those concerns. But I don't think we should be too apprehensive at the present time. One of the reasons why we merged with a company from California recently is because they understand the technology to reset this particular clock, so if it does prove to be a problem, we will be able to introduce that technology into the cloning process and we would then expect to be able to reset this particular clock in a perfectly appropriate way.

The implication was clear: Geron alone could solve this inherent problem that cloning did not reverse the aging clock with its use of the telomerase gene. My first response to the report was to ignore it. I didn't believe their results. The restoration of cell life span that Jose had reported was either true or fraudulent, and I knew Jose too well to believe the latter. Ian's data, on the other hand, wasn't convincing to me. His paper had information on one animal cloned from an adult cell, Dolly. Normally in science, you like to see things repeated before you accept them as fact. But the media loved the story, and soon the newspapers were full of stories about Dolly's premature aging.

My old colleague Jerry Shay even chimed in: "I recall when the news first came out, somebody said that Dolly was a sheep in lamb's clothing. I think that's an appropriate quote now."

A sheep in lamb's clothing?

One day on a taxi ride from New York City to La Guardia Airport, I struck up a conversation with my driver.

"So what kind of business are you in?" he asked.

"Biotechnology," I said, looking out the window.

"Well, what do you think about poor Dolly, being born old?" he asked.

"Where did you hear that?" I asked.

"Oh, I read it in a paper in Bombay."

Everyone on the planet, it seemed, had heard this story. I just had to prove it wrong.

When I got back to ACT I sat down with a new and brilliant employee, Bob Lanza. Bob was an M.D. and had a background in cell therapy. He was also a high-energy visionary and fearless. We decided on a stringent test that would settle who was right. We would grow cow cells to actual senescence and use them in a series of cloning experiments to test whether the cells remained old or were rejuvenated. Bob leapt into action, and within a week our experiments were under way. Having few resources due to our heavy commitments to supply cloned cattle to customers, we decided to lean on some outside collaborators. Peter Lansdorp from the Terry Fox Laboratory in Vancouver said he would help with some careful telomere length measurements, and Vince Cristofalo, who along with Len Hayflick was one of the original cell-aging scientists, agreed to measure some markers of cellular aging. We grew cells to senescence, and with Peter and Vince's help, carefully documented that the cells were indeed senescent and that they had short telomeres. Then from these old cells we did nuclear transfer. One day, after several months' worth of work, Bob Lanza came into my office excited.

"It is starting to look like this really does work," he said.

"Look, the telomere length seems to have increased after nuclear transfer."

Together we looked at the picture.

"Wait a minute, this is odd," I said. "Look, here are the original young cells and here are the old ones, and the old ones have decreased telomere length, as you would expect. But the animal cloned from senescent cells has telomere length longer than the original young cells. Maybe that's just a fluke, but in any event, the telomeres got reset. Those cells look young again!"

Later, Vince Cristofalo e-mailed us his results and Bob again knocked excitedly on my door.

"Mike, Vince is seeing the same thing! Using his own markers, the animals cloned from senescent cells look younger than young!"

We both knew we had to scramble to get a paper together. Not only had we shown the opposite result of the Dolly paper, we had for the first time shown the reversal of the aging of a mammalian cell. In the telomerase study we had not shown the reversal of the aging of cells; we had permanently immortalized them. Using nuclear transfer, we had reset the clock of cellular aging back to the place (or before the place) where life normally begins. And we hadn't done only that, we had made it embryonic again, capable of forming ES cells. Only one piece of data was missing to make a great paper. The telomere length and youthfulness of the cells had been restored, but how? Presumably it was through telomerase. But we hadn't provided any evidence that telomerase had in fact been turned on in the old mortal cell after it was transplanted into the egg cell time machine. We needed to provide direct evidence that we could awaken the telomerase gene through nuclear transfer.

Karen came to the rescue. She organized a two-pronged experiment whereby cow egg cells were fertilized normally by sperm to make embryos. In parallel, we had old somatic cells and embryos cloned from them by nuclear transfer. She then did the TRAP measurement, looking for the footprints of telomerase. The results were fascinating. Telomerase awakens in the normal sexual embryo shortly before the blastocyst forms, apparently to "check and make sure" the age of the cells is appropriately young, and to rewind the clock if necessary. In the cloned embryos, even though we had put an old mortal cell in the egg cell, the results showed that telomerase was turned on to levels comparable to those of a normal embryo!

Presumably the molecular Fates in the egg cell were "looking" at the shortest telomeres and in effect saying, "This won't do, the baby calf needs to be born young. Let's rewind the clock and add a lot more telomeric repeats." Telomerase, like the hand of Clotho, was weaving the thread of life and resetting the length back to normal. But in nature it would likely be unusual for the telomere length to be so short in the embryo. Perhaps this abnormal beginning played some role in overextension of telomere length.

The final data on actual cell life span came in and again verified our results. The cells from the clone were not only dividing, but they exceeded the life span of the original cells.

Can I possibly relate how fulfilling this experiment was? I believe I was witnessing the birth of a technology that one day would be used to make old cells young again, to alleviate the suffering of the aged, perhaps adding life to years as well as years to life. It was a means of making *zoë*, immortal germ-line cells identical to the patient. Philosophically speaking, it was a means of taking old cells back into the "womb" to make them born again. Recognizing the profound nature of the results, Bob Lanza named the animals after the symbols of the immortal renewal of life in spring: Persephone, Crocus, Hyacinth, Rose, Daffodil, Forsythia, and Lily.

At about that time, one of my old friends from Geron happened to be in town. All the employees of Geron had been instructed specifically not to have any contact with me, but that didn't deter the person in question from coming to our home for dinner. Afterward, I couldn't resist pulling out a draft of our paper. "Look at our data!" I said excitedly. I waited for him to scan the figures and digest the results.

"Wow, Mike, this is great!"

"Would you have expected this?" I asked. (It was my way of testing the water to see if Geron knew they had gotten the Dolly story wrong.)

"No, this is amazing!"

I found his response encouraging. It suggested that no one else knew that cloning reversed cell aging. That would help us get a patent on the technique. Geron's telomerase patent was irrelevant, because we weren't using the isolated telomerase gene.

We then wrote up a paper for submission to *Science,* which, a few months later, agreed to publish it. I tried again to write a measured press release, although it was difficult for me to restrain my obvious enthusiasm over the results. But then *Science,* in their own press release, said, "Researchers may have unlocked a cellular fountain of youth," and there we were, stuck again with that immortal metaphor.

Our paper was set to be published on my birthday, April 28, 2000. I was forty-seven years old. In the paper we showed a picture of six cattle cloned from senescent cells. Again, the stories written had a sensational edge. "Is Cloning Key to Fountain of

Youth?" asked MSNBC. "Cloning could offer a key to the fountain of youth, say scientists whose six cloned cows show signs of being younger than their chronological ages. Their cloning method appears to have reversed the aging process of cells, opening the door to creating organs that are nearly immortal for use in transplants."

In a story written by the Associated Press, Jerry Shay was quoted: "It's important not to overdramatize this as a 'fountain of youth' thing." Instead, he advised that we had provided "the first very dramatic proof" that the aging process could be reversed for medical purposes.

Perhaps a cartoonist in the *Tampa Tribune* said it best. His picture showed two young scientists at a lab bench with pictures of cloned cows taped to the wall. At the top of the comic it said, "Scientists Say Aging Process Reversed in Cloned Cows." An aging senior scientist enters the lab and lectures the young scientists: "I hope you guys do something to prove it. . . . I'd hate to see your long careers in research go down the drain."

In a commentary in *Nature Biotechnology* in June 2000, Ian Wilmut and Cal Harley fired back, discussing our paper and suggesting why we might have gotten things wrong. Science is like that—a competitive enterprise. We expected it. And after all, there are a lot of mistakes published even in good scientific journals. Ian was saying one thing, we another, so who was right?

Finally, in September 2000, Teru Wakayama, the scientist who had first cloned the mouse, gave an answer. After cloning one mouse, Cumulina (so named because she was cloned from a "cumulus" cell), he then cloned Cumulina herself, and then he cloned the clone of the clone, and so on for six generations in total. Teru published his results in *Nature*, reporting that, as in our study, telomeres did not shorten in the clones. "In fact," he wrote, "our results show that the telomeres lengthen with each generation."

Later, Xiangzhong Yang of the University of Connecticut reported that his cloned cattle had normal telomere lengths as well. And then finally, at a scientific convention, another anonymous Geron scientist leaned over and told me that Geron and Roslin were now seeing normal telomere lengths in their cloned sheep.

How do you spell relief? T-T-A-G-G-G.

But none of our work with animals would have full meaning until we could do the same thing with human cells. That was the next step, and to carry it out, we needed access to human egg cells. This would require a groundbreaking program in which women donated the cells specifically for the cloning of human embryos. There were all kinds of issues with this. We had to ensure that the women fully understood what they were doing, and more important, that they would be safe. What should be the standards for such a research program, and how would it be managed?

For the earlier work at Geron, I had organized an Ethics Advisory Board, a group of independent consultants to help us sort through the complex issues. Private companies have no legal obligation to set up such a committee, but the standard I wanted to apply was that every move we made should be absolutely beyond reproach. There were too many lives on the line. I could never live with myself if these new therapies were delayed because of my mistakes. If an EAB was justified before, then it certainly was here.

I knew of an expert in this field: Ronald Green, the chairman of the Department of Religion and director of the Ethics Institute at Dartmouth College. Ron had authored a scholarly work on the history of this controversial field titled *The Embryo Research Debates: Bioethics in the Vortex of Controversy.* He graciously agreed to donate his time and worked hard to organize an ethics advisory board for our company. Over the following months, he assembled a cadre of thoughtful people to begin to debate all of the issues, such as how much we would compensate women for their time and trouble, and how we would ensure that they were completely informed as to medical risks and the nature of the research in which their cells would be used. We needed to decide on the extent of medical examination necessary to ensure the lowest possible risk to their health, and we settled on a level of hormones sufficient to yield a few egg cells but low enough to be far safer than that normally used. Finally, after what felt like an eternity, the board gave us the green light to proceed.

Ann Kiessling, director of the Bedford Research Foundation, whom we had recruited to become part of the experiments, arranged with an experienced medical team to coordinate the egg cell donors, and in the spring of 2001 the first human egg cells came to ACT. To no one's great surprise, our first try was a complete failure. Not a single egg cell survived the procedure. But as the summer of 2001 wore on, we steadily improved our technique. Finally, in one experiment we looked at the embryos and saw the first evidence of success.

When the sperm cell enters the egg, and if everything is operating normally, the egg cell will "unwrap" the DNA from the sperm and program it to be in an embryonic state. In cloning, if one sees this reprogramming, as evidenced by the formation of a large and unusual-appearing structure called a *pronucleus,* then there is reason to believe that the somatic cell has been restored to an embryonic state. Most important, I believed that the appearance of the pronucleus was evidence that telomerase was at work, rejuvenating the cells.

However, as has so often happened in attempts to clone a new species, the cell would not divide. We were confident that we were very close, but our goal was to make stem cells, so we pressed on.

During that summer, as we worked on human embryo cloning, we watched the heated debate unfold on the issue of federal funding for embryonic stem cell research. On two separate occasions we had waded into the debate ourselves by drafting letters to the government urging Congress to allow funding through the National Institutes of Health. Bob Lanza happened to have the fax numbers of nearly a hundred Nobel laureates. We faxed off our letter to every one to see if they would add their names to the cause. We were grateful to receive back dozens of signatures to add to our own. These letters made an impact, though opponents were quick to respond with the criticism "Well, they didn't get the Nobel Prize for ethics!" I found this remark somewhat disingenuous. The Church was quick to proclaim itself an authority on both ethics and science, with the Pope calling for the greater use of adult stem cells to cure disease. If the Church can pontificate on matters of science and medicine, then I think it is entirely proper for members of the scientific community to voice their position on matters of medical ethics. I think it is not only appropriate, but mandatory, that scientists be heard.

Then, on the evening of August 9, 2001, I heard that the President, in his first national address, would tell us his decision on the matter. I sat in front of the television watching this surreal event unfold before my eyes. This microscopic project, till recently of interest to only a handful of scientists in the whole world, was now the subject of a major presidential address. Virtually the whole world now knew what embryonic stem cells were. That night, in eleven minutes, George W. Bush told the American people how he had agonized over the issue. On the one hand, he believed in the sanctity of human life and the preimplantation embryo was a human life. On the other hand, human ES cells had the promise to change the face of medicine and save the lives of millions. In a Solomon-like manner, he then announced that he had decided on a compromise. He would allow federal funding on cell lines created and already in existence as of August 9, 2001. Any made after that date would not be eligible for funding.

On the face of it, this seemed like a good solution. But there is often a problem when decisions are made based on anything other than solid facts and rationality. While the President's position would allow some research to proceed that could in fact produce a miracle in the laboratory dish, the corresponding miracle in the hospital bed (an actual cure of a disease) would not be possible. The cell lines that existed before August 9, 2001, were far too few to make a library of cell lines large enough to match up the many millions of tissue types in the human population. We would need a library of at least a hundred thousand ES cells in the future to be able to ensure that every patient in need of a liver or a kidney transplant could have even a close match. Even then, because the stem cells would not be based on an individual's own DNA, future patients would need to take drugs to prevent rejection of the alien cells. *What we needed was to advance therapeutic cloning!*

Of course, the die-hard opponents weren't happy anyway. They wanted all ES cell research stopped. Indeed, the Christian Legal Society filed suit challenging Bush's authority to issue such funding in light of the Dickey-Wicker Amendment. Passed in

1996, the measure banned any federal funding being used for research in which human embryos were destroyed.

About this time, a reporter with *The Wall Street Journal* called me on the phone.

Antonio Regalado is more of a detective than a reporter. He had covered the ES cell debate in numerous previous articles, often presenting the scientists working in the field as shadowy figures, peering from behind closed laboratory doors.

We had been so careful about our donors, and our procedures were highly confidential. We never knew, ourselves, the identity of any of the donors and had put in place measures to ensure that we couldn't find out even if we wanted to. Nevertheless, Regalado had searched through the classified ads in the Boston newspapers where we had recruited egg cell donors. But the worst part about his article was the headline, "Ethicists, Bodyguards Monitor Scientist's Effort to Create Copy of Human Embryo." It implied that our work was so controversial that we needed bodyguards to protect us. The truth was that our Ethics Advisory Board wanted the bodyguard to oversee the transport of the egg cells and to make sure none could be diverted for anything other than the use we had in mind. In other words, the bodyguard was for the egg cells, not the ACT scientists.

The article concluded with the statement: "Somebody is probably going to get shot."

That quote hurt us enormously. One of our investors called me on the phone, backing out of our financing.

"Mike, we just can't be associated with such controversy," he said. "Maybe we are being wimps, but that is just the way it is."

I sensed that we were heading for even stormier days. We had other strategies under way at ACT that were also very controversial. In an effort to get totipotent stem cells from other places in the germ line, we took monkey egg cells and found a way to start them growing as embryos on their own, without fertilizing them with sperm or by using nuclear transfer. To do this, we "tricked" the egg cell into thinking it was fertilized by a sperm. The egg cell then began dividing into two cells, then four, and within a week it became a blastocyst. From the blastocyst we derived an ES cell line. From that line we made teratomas with skin and hair, intestine, beating hearts, and neurons useful for treating Parkinson's disease. In other words, we had turned an egg cell into beating heart cells.

For a woman of reproductive age, this would allow us to make many medically useful cell types, and as in the case of cloning, all of the DNA came from the patient, so we would not expect rejection to occur. Because the egg cell does this on its own, it is called *parthenogenesis* (meaning "virgin birth" in Greek) and the embryos made in this manner are called *parthenotes*.

Parthenogenesis works perfectly well in several species. There are lizards that use only parthenogenesis to reproduce; there are no males, and each offspring is a "clone" of the mother. And the common dandelion reproduces parthenogenetically; the dandelion flower makes pollen, but it is not used in reproduction. However, in the case of mammals, embryos made parthenogenetically and then transferred into a uterus can develop as fetuses for a period of time, but they always develop abnormally and die before birth. Nevertheless, the cells made from parthenote embryos may be normal enough for many medical purposes.

But the Roman Catholic Church was already upset with our proposals about cloned human embryos. What were they going to say about parthenogenesis? Theologians had always dominated the province of the virgin birth. But now we knew that these types of embryos had several unique properties. I was certain that laboratory incubators around the world would likely soon be brimming with them. I could already imagine the invective that would be thrown at us from those who venerated the virgin birth of Christ.

The tragic events of September 11, 2001, paradoxically gave us a brief respite from the media onslaught. The attention of the entire nation was riveted by the cataclysmic terrorist threats to our security. But a month later, by the night of October 9, when we ran our experiment in the human-cloning lab described at the start of this book, all of us at ACT were feeling the pressure intensely. Given a hostile White House, sensation-hungry media, and the ever-present financial strain, exacerbated by the generally dismal economic climate, we experienced an exhausted, almost now-or-never desperation when we finally broke up for the night. It was as though our very fate depended on the mysterious molecular processes that were or weren't taking place in the darkness of that incubator.

On the morning of October 10, Jose returned to the company. He had gotten too little sleep, but he couldn't stay away. Like all scientists passionately involved in their research, he simply had to look one more time at the results of our work the night before.

He unlocked the incubator and slipped out the small plastic dish containing the seven cloned human embryos. He switched on the microscope and focused carefully on the nuclei.

There, embedded within the egg cells, were unmistakably the large and unique forms of pronuclei. The DNA looked reprogrammed. Excellent!

But it was only a small first step.

Again, nothing to do but wait.

Two days later, on October 12, Jose again opened the incuba-tor. He was expecting the worst.

Shortly afterward, there was a knock at my door.

The normally unflappable Dr. Cibelli was short of breath.

"Mike," he managed finally, "you have to come see this."

I followed him back to the human cloning lab. The door slammed behind us. Jose said nothing; he just stared at the mon-itor over the lab bench, a big grin from ear to ear. There on the black-and-white monitor were three dividing embryos.

Oh my God!

Again, there was nothing to do but wait.

Four days, five days.

By day six, the cells had stopped dividing. Six cells at most, and that was it. We were far, far short of the approximately one hundred required to make the blastocyst that was our goal. But at the same time, we'd gotten to pronuclei and cell division. This was far ahead of what we—or anyone else—had ever achieved.

Along the way, though, I got another call from Regalado of the *Wall Street Journal.*

"Should I say you have succeeded in making stem cells from cloned embryos?" he asked me for openers.

Good grief, had he heard something? How could he have known one way or the other? I decided on the spot that he couldn't—that he was bluffing: "Antonio, we really plan to describe everything we're doing at the proper time, in a scienti-fic publication, and that is about all I can say for the moment."

"Should I take that to mean that you have the data and are writing up a paper?"

"Give me a break, Antonio. You know this is serious research. We will release our data at the right time and the right place."

End of conversation.

Our intent had been to quietly get the cloned human embryos to the blastocyst stage, produce our first stem cells, and only then publish a scientific paper that, we hoped, would rally the scientific community in support of our work. But now there was a risk of our being compromised. Who knew what Regalado would put into print?

Jose, Bob Lanza, and I gathered around in our conference room to decide what to do next.

There was an additional problem, too. The BBC would be airing some footage we'd permitted them to take of Jose performing human nuclear transfers. The program was scheduled for the following January, February at the outside. Possibly we could stall Regalado that long, and stonewall any other reporters who were on the story. But that seemed to us an outside limit, and if we did nothing, there was always a chance the public would learn what we were up to in the worst tabloid headlines.

There was too much at stake. We felt—unanimously—that we simply couldn't let that happen. The solution we agreed upon was far from ideal, but we all decided to publish a paper of our own quickly, sharing the data in our possession. We had a precedent for this approach. Groups working on monkeys had published papers on nothing more than a pronucleus. And Bob Edwards, in the course of making IVF a reality in the 1970s, had published papers every step of the way: when the sperm fertilized the egg, when he got cleavage, and so on. For our part, we had three things to report: pronuclear formation, cleavage of cloned embryos, and human parthenogenesis with some early embryonic development. So even though it was far earlier than we had planned, we thought it would be better if we gave the world a progress report ourselves than for a Regalado to print God-knows-what and bring the media and the crazies banging on our doors.

At the same time, a reporter named Joannie Fischer with *U.S. News & World Report* was in the midst of an in-depth story on ACT. She was camped out at the hotel that formed part of the biotech complex containing our offices, and she was spending days at a time with our scientists. While our key experiment had been conducted at night, Joannie was no fool, so it was no surprise when one day she walked into my office with some leading, and quite detailed, questions. I liked Joannie, who clearly understood the balancing act of producing good journalism without pouring oil on the flames of public controversy. Once she had assured me that she would agree not to print a story until we ourselves had published our scientific paper, we decided to give her free access to everything we were doing.

But Joannie, it turned out, couldn't speak for all of *U.S. News*. The next thing I heard, they wanted to make her reportage a cover story, and she was having a hard time getting management to agree to hold off. I flew to Washington to meet with Joannie's bosses and plead our case. My message was simple: *U.S. News* could print anything they wanted—they could demonize us, they could praise us as saints—but to do so before we had published our paper would be plainly unethical. In the end, they agreed, but they would not give us a hint of whether their coverage would be favorable or unfavorable. The one thing they did show me, though, was a paste-up of the photos they would use of Jose, Bob, and me. Under the picture was the caption "The renegade mad scientists."

"Hey, that's not the actual caption, is it?" I asked the layout man, breaking into a sweat. "The one you're going to use?"

He laughed.

"We're entitled to some fun too," he said. "No, that's just a place filler, don't sweat it."

Hey, who was sweating?

Our paper showing our cloned human embryos was scheduled to be released on the Internet by *e-biomed: The Journal of Regenerative Medicine* on Sunday, November 25, 2001. On the Friday before, Joannie called to say that *U.S. News* normally distributed its magazine to Capitol Hill on Saturday, but they were thinking of holding it. They had arranged for her to be on *Meet the Press* on Sunday. They thought they would break it there.

"Mike, it would be wrong for you not to be there as well. *Meet the Press* will also show the cover of the January *Scientific American*, but they [*Scientific American*'s people] won't be there."

Bob, Jose, and I had been writing a review on human therapeutic cloning for *Scientific American,* and when *U.S. News* decided to do their cover story, *Scientific American* decided to do the same and to release their story a month early, so as to appear on the same Sunday.

"We think you should be there," Joannie repeated.

"Well, why not? I don't have any plans for Sunday, do I?"

As I waited in the green room of *Meet the Press* in their Washington, D.C., offices that Sunday morning, a representative of *U.S. News* at last handed me a copy of their article. By this point, I was prepared to defend our work against the most bitter of attacks, but as I skimmed the pages, their story seemed fairly balanced. The caption under our picture read simply, "The Cloning Troika." The cover, however, was troublingly provocative.

"THE FIRST HUMAN CLONE," it said in bold capital letters.

I walked into the studio and wired up for the interview around a modern-looking blond wood table.

"It's a pleasure to have you here," said Tim Russert, shaking my hand warmly. "It looks like you've really started something here with this whole cloning thing."

"Well, it's something we desperately believe in," I answered humbly.

Maybe I should have been jittery. I, a relatively obscure scientist, was about to be on *Meet the Press* and blithely tell the whole world we had cloned the first human embryos. What in the world had I gotten myself into?

But for whatever reason, I was quite calm. The only things on my mind were the lives of the people we could help and a desperate desire to head off the stampede to legislate away this whole technology. Maybe the stampede would come anyway, but knowing in my heart that I was on the right side of the debate gave me confidence. This was my chance to give it my best shot.

Within seconds came the countdown, and suddenly we were on the air.

Tim Russert turned to me and asked, "In layman's language, what have you done?"

"Well," I answered, "we've taken the first halting steps toward what we think is going to be a new era of medicine. It's been called regenerative medicine."

I then explained the difference between reproductive cloning to make babies and therapeutic cloning, which I said is designed to "take a patient's cell back in time using the egg cell as sort of a little time machine, and then making these cells that we've heard so much about over the last few months, the embryonic stem cell."

Bob Bazell, NBC's science correspondent, pointed out that our publishing the recipe for human embryo cloning would make it possible for another group to use the identical technique to clone a child.

I responded in measured tones, simply saying we'd considered that possibility carefully. But there was already so much published on the subject, we doubted we'd given away any secrets of significance.

My segment of the program over, I headed toward the door, only to bump into a female producer from CNN. She seized my arm, as though to make sure nobody else could whisk me away.

"Are you available to go right now to our studio?" she asked.

In for a penny, in for a pound. As we left the studio through the main entrance, a flock of reporters were waiting with cameras running and microphones poised. I assumed that they did this to interview all those who had just been on *Meet the Press*. I answered a few quick questions and then was rushed over to the CNN limo. That day launched me on a spate of such interviews.

It was late that Sunday night before I got back to Massachusetts, to Karen and a house with two-month-old triplets, and took my turn with the midnight-to-six feeding shift. With a baby feeding in each arm and a pencil in my mouth, I scrolled through hundreds of e-mails that had come in that day. Well over 95 percent, then and later, were encouraging us to work even harder. They wanted cures for their child in heart failure, they were volunteering for clinical trials for Parkinson's—whatever affliction it was, they were reaching out.

The following day, however, the newspaper stories took the line that we had "shocked the world," and the attacks were quick to follow. Our paper and all the publicity had taken the opposition by surprise, but now they came out with guns blazing. On Monday, November 26, President Bush, welcoming back the aid workers rescued from Afghanistan, took questions from reporters in the Rose Garden.

"Mr. President, it seems an important line has been crossed with the attempt to clone a human being by a private laboratory. What's your reaction to that? Do you think there's any way to put this genie back in the bottle?"

The President replied, "My reaction . . . is that the use of embryos to clone is wrong. We should not as a society grow life to destroy it. And that's exactly what's taking place. And I have made that position very clear. I haven't changed my mind. And this evidence today that they're trying to achieve that objective, to grow an embryo in order to extract a stem cell, in order for that embryo to die, is bad public policy. Not only that, it's morally wrong, in my opinion."

Later in the day, at a press briefing by Ari Fleischer, came further questions. Specifically, in light of the President's opposition to our work, did he plan any steps against us?

"Does he think that doctors who engage in the kind of research that these people in Massachusetts do should—should be disciplined, sent to jail, what?" a reported asked.

"The President thinks that the practice of cloning should be made illegal," Fleischer answered.

"Then the question is, what do you do with the people who do it?"

"I would refer you to what the penalty section was of the House bill. I don't have the bill in front of me. But he has supported the Weldon legislation."

The Weldon legislation Fleischer referred to was a bill that had already been passed by the U.S. House of Representatives on

July 31, 2001. If it had become law—that is, if the Senate had also passed it and the President signed it—it would have fined any scientist transferring a human somatic cell into an oocyte $1 million and sentenced him to a ten-year prison sentence. Besides being a horrible mistake for the future of medicine, it would have been a terrible precedent on government intervention in basic scientific research, not unlike locking Galileo in prison for simply turning his telescope to the stars.

Some legislators, meanwhile, saw far more ominous implications in our work. "We are on the verge of having human embryo farms in laboratories all across America," said Chris Smith (Rep., NJ). He called our report "ghoulish work."

On that same Monday, the Vatican stated that our research "moves us to restate with force that the beginning of human life cannot be fixed by convention at a certain stage of embryonic development; it takes place, in reality, already at the first instant of the embryo itself." And it added: "Thus, despite the declared 'humanistic' intentions by those who predict sensational cures via this path . . . what is needed is a calm but firm judgment which shows the moral gravity of this plan and which motivates an unequivocal condemnation."

The Orthodox Church stated, "The human cloning experiment announced in the United States brings to mind the crimes against humanity of a Nazi brand . . . they are sorcerer's apprentices who don't know where they will end up. It is a defiance against God that will lead to grave disasters."

Two days later, on November 28, the President signed an executive order establishing a new President's Council on Bioethics, to replace the National Bioethics Advisory Commission established by President Clinton, and about a month later Bush appointed Dr. Leon Kass to head it. Kass had been one of the leading opponents of in vitro fertilization (IVF) in the late 1970s. He called it the "dehumanization" of reproduction.

I can no more see how a rational person could object to IVF than I can understand the objection to the use of cloning in medicine. But with the conservative appointments to the new council, many of whom were among the most vocal opponents of the technology, it was no surprise to hear that they would recommend legislation to stop it.

In the months that followed, there were numerous hearings in the Congress on what laws the United States should pass to regulate human cloning.

Nigel Cameron, director of the Council for Biotechnology Policy, said, "ACT, under cover of its secret ethics committee, has forged ahead. While we have been focusing on the prosecution of the war, they have sought to shift the conversation by a *fait accompli*. And while we were celebrating Thanksgiving with our families, through a slick PR maneuver Michael West has made their announcement to the world. They must be stopped."

Another group to condemn us was the Christian Legal Society. Within twenty-four hours they had asked the Commonwealth of Massachusetts to prosecute us for having violated one of the Commonwealth's statutes banning research on live aborted fetuses.

Meanwhile, the Department of Commerce announced that ACT had been awarded a sizable grant for work unrelated to human cloning. Shortly thereafter, a member of the House of Representatives, Joseph Pitts of Pennsylvania, called for an investigation of ACT to see if it had used any federal grant monies in support of its human embryo research. In 1996, the Dickey-Wicker Amendment was passed; it barred any entity from using federal money in research that resulted in the destruction of embryos.

The Office of the Inspector General then notified us that we were to be audited.

One day, as Karen and I were on our way out the door of ACT to have lunch, three ominous-looking suits were heading in toward our offices.

"They must be the OIG auditors," I said, for they certainly looked like G-men.

As far as I knew, we had never used any government funds in the human embryo work, and I gave the auditors free rein. The next thing I knew, threats of criminal charges for fraud were in the air.

"What's going on in there?" I asked one of our scientists who had just come out from being interviewed in our conference room.

"They're on the phone with Tommy Thompson, and he just told them to nail us for the fraudulent misappropriation of grant funds."

Tommy Thompson? He was a member of the President's cabinet. That got my attention.

"But we haven't used any federal money for the human embryo work. It's impossible."

"They're saying Teru bought a microscope that wasn't in the budget."

"Teru" was Teru Wakayama, the scientist who had cloned the first mouse.

"You've got to be kidding. A member of the President's cabinet is on the phone about a *microscope* for a *mouse-cloning* project? Doesn't he have anything better to do?"

Besides, we had approval for the microscope, and I could prove it! But that wasn't the point. We were in a witch hunt, pure and simple, a plan to make us look bad, and they were going to do just that regardless of the facts.

The official report of the OIG was released on April 26, 2002. It stated that "we found no evidence that NIH funds supported ACT's human embryo cloning activities." Despite the fact that the microscope was an approved purchase, they asked that we return the funds since it wasn't in the original budget. The OIG report summarized that "our report does not state, nor do we imply, that ACT misappropriated grant funds to purchase the equipment."

Should it have surprised me, then, that Representative Pitts would thereupon issue a press release saying that "HHS says Advanced Cell Technology misappropriated funds" and that "ACT may have used taxpayer funds to subsidize its human cloning activities"?

On the same day, Representative Pitts and Curt Weldon (author of the Weldon bill in the House) wrote Tommy Thompson, saying, "As the audit points out, ACT has misappropriated taxpayer funds."

The newspaper headlines were incriminating and frankly upsetting. I called my attorney. "Isn't this an abuse of power?" I asked. "A member of Congress could bash any U.S. citizen in this manner and destroy their career."

"Mike, they have immunity from prosecution for libel. They do this all the time."

Some members of the religious community, instead of taking the high road by calling for a rational debate, also resorted to inflammatory rhetoric. Richard Land of the Southern Baptist

Convention called therapeutic cloning "high-tech cannibalism in which we consume our young."

But perhaps the nadir of the rhetoric was an article in *The Weekly Standard* by William Kristol and Eric Cohen, which ran with the headline:

Dr. West and Mr. Bin Laden:
Cloning and Terrorism Are Both Clear
and Present Dangers

In the meantime, the company was running out of money, and I was running out of ideas for how to meet next week's payroll.

On December 4, I was asked to appear before a Senate committee and be grilled one more time. I struggled through it, tired and battle-weary. The next day, back in Massachusetts, I took a long walk, trying to figure out where I'd been, where I was going. I remembered that day I had sat across the street from my hometown cemetery and my all-out determination to fight the battle I had since fought. I knew back then that it would be hard, but this was harder than I had expected.

At the same time, words of advice Sam Goldstein had once given me echoed again in my ears: "Mike, it's like a boxing match. They knock you down, and you have to get up and keep on punching."

Back home, I sat down to sort through hundreds of unread e-mails. They touched me to the core. I happened on one from a man with a spinal cord injury. "What you are doing is revolutionary. You are the one in the arena, and the timid few on the sidelines, most do not know the harsh reality of chronic paralysis. Senator Brownback cannot dictate medical breakthroughs as right or wrong. What you have started could very well end my suffering. There are always going to be the doubters, yet by quickly trying to get therapies to the bedside, you and your company are unselfish, innovators, geniuses, and most of all offering hope to the thousands who have been told to just deal with their physical, degenerative disabilities. You are trying to do good, period! Anyone else who sees differently [has] not suffered or had a loved one suffer."

This correspondent was not alone. Like the slow lifting of storm clouds that leads to the warmth of the sun, the messages renewed my strength.

Somehow we managed to weather the storm, even when it meant braving the right-to-life picketers who gathered in the street outside our building. I didn't have to tell my colleagues that we had work to do. We were all determined to forge ahead. If anything, the state-of-siege war mentality, with Congress poised to act but not acting, lent an urgency to the work, and in the never-never land that followed, with our finances teetering and fanatics of every description demanding our scalps, we persevered.

Chapter Eight

The Abolition of Death

Then the LORD God said, "The people have become as we are,
knowing everything, both good and evil. What if they eat
the fruit of the tree of life? Then they will live forever!"
So the LORD God banished Adam and his wife from the
Garden of Eden, and . . . the LORD God stationed mighty angelic
beings to the east of Eden. And a flaming sword
flashed back and forth, guarding the way to the tree of life.
—GENESIS 3:23–24

And which of you by taking thought can add one cubit
unto the measure of his life?
—JESUS (Matthew 6:27)

How would you summarize in just one sentence what it means to be a human being? In reality our lives are so complex. Every day is brimming with the clutter of details: the mortgage on a new house, fitting our children in a new wardrobe of summer clothes, and so on. Few of us ever take the time to think through what some call "the meaning of life," let alone try to distill down the essence of life itself into a few words.

In the late 1960s, Jon Lomberg and Carl Sagan offered to do just that. They urged NASA to consider bolting a small gold plaque to the *Pioneer 10* spacecraft, which was destined to fly past the planet Jupiter. This was a flyby mission, so after the spacecraft finished its job of photographing the largest of the planets, it would be aimed off into the depths of the universe without any further purpose. Lomberg and Sagan hoped to use this spacecraft to transport a plaque into the infinity of space, to whatever intelligent life may find it millions of years from now, describing who we, the people of the Earth in the twentieth century, were.

In the closing years of the 1960s, as the counterculture listened to the Beatles' "A Day in the Life," Lomberg and Sagan made a simple line drawing of a naked man and woman and mailed it off to NASA. They reasoned that from the etching an intelligent being could conclude that we were living beings who reproduced sexually. This would imply that we evolved as a mortal soma. The man was drawn with his right hand raised in a gesture of friendship. From this depiction, it was hoped that another life-form would understand that we were capable of not only cognition but also of compassion. After all, I think that we are a species that is proud of the fact that we are capable of love, not just the romantic attachment of two people, but love in the more classical sense, what the Christians call *agape*, a love characterized by the wish to benefit others, even at personal expense.

Lomberg and Sagan communicated in this small plaque the essence of what it is to be human. We are capable of love, and we stand to lose that love through death. This tension is as central to the human condition in our own generation as it has been from the beginning of civilization.

The ancient Babylonian *Epic of Gilgamesh,* written about 2500 B.C., is the first recorded account of the battle of love against death and the fear of arousing the animosity of the gods in trying to change the order of nature. Gilgamesh was a powerful figure who learned of the certainty of death and set out in a quest to find the green plant called the *"shibu issahir amelu,"* which bestowed an immortal renewal of life. Against the will of the gods, he valiantly fought to find the plant. He eventually succeeded, but in a moment of weakness, he stopped for a drink of water and set the plant down by his side. A snake came by and ate it, thereby conferring immortality to the snake, but destining humans to the certainty of aging and death.

In our day, the same quest is being played out, not in cuneiform and clay tablets, but in the laboratories and in the governments of the world. However, the weapons at our disposal are vastly more powerful than anything imaginable in the ancient world. For the first time in history, we have the potential to transfer the immortality of the germ line into the soma, along with other powerful new insights into the biology of aging and disease that have the potential to build a new and brighter world only dreamed about in ancient Mesopotamia.

But, critics will carp, "like the *Epic of Gilgamesh,* the gods are not on our side in this quest." One such critic, the British historian Paul Johnson, wrote in *The Spectator:* "In the 20th century, we failed to stifle at birth the totalitarian concepts which created Nazism and Communism though we knew all along that both were morally evil—because decent men and women did not speak out in time. . . . Are we going to make the same mistake with this new infant monster [biotechnology] in our midst, still puny as yet but liable, all too soon, to grow gigantic and overwhelm us?"

As a person who has labored to give birth to this "infant monster," I take these criticisms to heart, and have struggled to understand the concerns expressed by those who oppose us in this debate. I understand the fears we all have of the unknown. In these opening years of the twenty-first century, our nerves have been frayed by horrific displays of terrorism and inhumanity. But disease is terrifying as well. Just as rescue workers were courageous enough to risk their lives to save their fellow human beings in the rubble of the World Trade Center on September 11, 2001, so we all need to display courage in applying modern medical technology with the hand of compassion. In this respect, I actually value the title of Aldous Huxley's novel *Brave New World,* for we need to make a new world—not the world of Aldous Huxley's novel, but a world free of the scourges of diabetes and heart disease.

Ethical discussions to discern the good and the bad uses of technology are often divided into two categories. First, the deontological perspective, from the Greek word *deon,* meaning "duty," focuses on our duty to God. It is generally religious in nature. A second is the teleological perspective, that is, do the ends justify the means? In this case, does the fact that the ends, or telomeres, of chromosomes are repaired by cloning justify the means (creating a preimplantation embryo)?

Like the perennial opposition of two political parties, those whose religious beliefs set them against such research argue with those from the medical camp. In the religious camp, the debate over therapeutic cloning hinges almost entirely on the issue of when a human life begins. This, I think, is the easiest point to consider, so let's begin with it.

The Sixth Commandment says, "Thou shalt not kill." Is the destruction of a preimplantation embryo murder? I believe that the answer is no, for it is not the taking of an actual human life.

First, before I explain why I think this is so, it would be helpful if we all admit that our gods have given us no clear instructions on the matter. The Bible, Torah, and Koran were not written in the era of molecular and cellular biology. Anyone familiar with the world's religions will, I suspect, quickly acknowledge that the ancient dogmas provide little of concrete use in formulating modern standards of ethics and law relating to the preimplantation embryo. This should engender in the truth-loving person a high level of caution. As a person wishing to please God, how can you be certain that God is on your side? For example, where leaders of one world religion take a stand against ES cell and cloning technology, the leaders of others argue strongly on its behalf. Indeed, the Union of Orthodox Jewish Congregations of America, representing a thousand synagogues, announced that they supported human therapeutic cloning. Dr. Edward Reichman, an Orthodox rabbi at New York's Einstein College of Medicine, said, "[A] fertilized embryo in a petri dish does not have the status of human life." The lesson, I would argue, is that when God is silent, we must be cautious not to insert our own prejudice.

In using the egg cell as a time machine, we are not proposing the cloning of a developing embryo, a pregnancy, or a fetus. Instead, therapeutic cloning, as the term is generally used, refers to the transfer of a patient's somatic cell into an egg cell to form a "preimplantation embryo," from which one would derive an ES cell line specific to that patient. The point of disagreement is the moral status of such a preimplantation embryo.

There is a popular saying that "life begins at conception." That biology teaches us this popular aphorism is an unfortunate mistake. Certainly, "life" doesn't begin there. The sperm and egg cells are very much alive, and as we have considered in earlier chapters, the germ-line lineage of cells is not only life, it is immortal life. What I think people mean to suggest is that *a* human life begins with conception. But this is wrong as well. Quite often a single fertilized egg splits to make twins. Since they come from a single fertilized egg, they share all the same genes—they will be identical. Certainly in the case of these twins, no one can assert that they were individual human beings at conception.

In some mammals twinning always occurs; in humans only in about one in three hundred births. And depending on how the membranes enshroud the babies at birth, the obstetrician can estimate when the embryo split into two. If the twins share no membranes at all, then the embryo split early in the first week after conception (that is, somewhere between the two-cell and the 32-cell stage). If they share only one membrane, then the embryo split sometime toward the end of the first week after fertilization (or somewhere between the 32- and 200-cell stage). If they share two membranes, then it split sometime toward the end of the second week after conception (when it was between 200 and 1,000 cells). So when *did* they arise as separate individuals?

Scientists say that "individualization" occurs at about fourteen days after fertilization. Up to this point there has been no movement toward building a human body. The cells in pre-implantation embryos contain no body cells of any kind, not even any cells on their way to becoming body cells. They are blank embryonic and immortal stem cells. They are not different from the human embryonic stem cells grown in laboratories around the world. The metaphor the cell biologist would use to describe these cells is that they are like the raw lumber from which one would build a house. If a lumberyard burned down, no one would report the event as "twenty houses burn to the ground." All that burned down were raw materials for houses.

They were only *potential* houses. In the same way, scientists and ethicists generally agree that a ball of unformed cells like this is merely the raw material to make a human being, or two human beings, but it is not one yet.

Shortly after implantation, a woman can obtain a positive result on a pregnancy test. This is the time when the embryo has buried itself within its mother's womb and begun the process of development. The first such step occurs when a line appears on the blank cells. The line is called the *primitive streak*. It is a line drawn by nature, not unlike an artist's first stroke of paint on a canvas. If two primitive streaks form in parallel, then identical twins will result. If one primitive streak forms, one human will result. If two streaks form but converge like a Y, then conjoined twins will result—that is, twins that share a head, or two heads that share a body. I would argue that this, therefore, is the defining moment when *a* human being begins development. The primitive streak is therefore a very useful line drawn in the dust of the ground from which we are made, drawn not by arbitrary human convention but by nature itself.

Some would disagree with me and say that all this shows is that at least one person is present in the preimplantation embryo—maybe two, which would be even worse. But the reality is that occasionally, though rarely, two different embryos will stick together to make what is called a *chimera*. This is one person that was originally two embryos, each conceived from their own sperm and egg cells. If one of those embryos was female and one male, the chimera will be a mix of male and female cells. Sometimes this is manifested as a hermaphrodite, a person with both male and female genitalia. The presence of chimeras existing naturally in the real world again leads us to consider that *a* human life isn't a determined event at conception.

Another way of seeing the contingency of the preimplantation embryo is to consider the fact that, in normal sexual intercourse, some 60 percent of fertilized eggs and preimplantation
embryos never attach to the uterus and simply die. If these were
people with immortal souls, then heaven's streets of gold would
be covered with souls who never passed the 100-cell stage.

The preimplantation embryo may be destined in the natural
setting to be one individual, two or more individuals, no individual, or part of an individual. In other words, its fate is *contin-
gent*—it is a potential individual only. In that case, all sperm and
egg cells are *potential* individuals, and all cells in the body are
potential people through the power of cloning technology. But
the reason no Christian leader is yet calling for an eleventh commandment—"Thou shall not kill a skin cell"—is that although a
skin cell *could* become a person, it is *not* actually a person. Neither is the preimplantation embryo.

If the preimplantation embryo is not committed to being
one, two, four, or a fraction of an individual, it is not actually an
individual, and something that is not an actual individual
should not, in my opinion, be identified as a person by law with
a right to life. Every man makes millions of sperm and every
woman makes thousands of egg cells. No one would claim that
they should all be adopted and given a chance at real life.

Cells taken from any source, even a cadaver, always carry
their own weighty ethical issues. A preimplantation embryo has
no body cells of any kind, never has been a human being, and
has no sentience. Its potential for life is akin to the potential of
an acorn to become a tree. It is true that the mere event of set-
tling into the ground is enough to provide a home for it to
branch out into the wonder we call a tree, but still that does not
mean an acorn is a tree or that it should be treated as one.

Is it right to measure the value of cells that hold only the potential to be a human being, or two, or a part of a human being, with the same measuring stick that we measure a living and breathing human life? To borrow a phrase from Albert Einstein, I don't believe God plays dice with the human soul.

It is indeed tempting for some to argue that out of fear of offending God—that is, out of fear of making a mistake in judging our *duty* to God—we should bury these talents of gold and place a ban or moratorium on their use. But if history teaches us anything about these conflicts of science and religion, it is that vague fears alone should not be the driving force of our decisions. I would argue that courage should motivate us—courage to sift through the complex issues of right and wrong and diligently use the good to lift the burden of pain from those afflicted with disease.

The conservative right, such as J. Bottum writing in *The Weekly Standard*, weighs in on the side of abundant caution: "The immortality project, the perfect baby project, and the universal-happiness project are all aimed at the same end: the amelioration and consequent elimination of the human condition. Our notions of natural rights, our claims of human dignity and equality, are all based on the complex interplay of birth, health, aging, and death. And when these have changed as completely as biotechnology wants to change them, what will remain of rights, dignity, and equality? . . . Unless we embrace as a culture some coherent unmodernism, there is no preventing the biotech future."

History is replete with examples of where fear and superstition lead to "unmodernism." For example, the use of anesthesia in childbirth to ease the pain of labor is widespread and accepted today. But when it was first proposed, when the British obstetrician James Y. Simpson first tried chloroform in 1847, it generated fierce opposition from Church leaders. After all, there was a strong case to be made directly from the pages of the Bible itself that this was wrong. The Book of Genesis explicitly pronounced God's judgment on women for giving the forbidden fruit to man: "To the woman, He said, I will multiply your grief and your suffering in pregnancy and the pangs of childbearing; with spasms of distress you will bring forth children" (Gen. 3:16). One New England minister went so far as to preach, "Chloroform is a decoy of Satan, apparently offering itself to bless women, but in the end it will harden society and rob God of the deep earnest cries which arise in time of trouble, for help."

Other examples abound throughout our history. When smallpox inoculation was proposed in the eighteenth century, it was denounced as evil by some of the leaders in the New England Church. Increase Mather relates the thoughts of William Cooper, a Cambridge pastor: "The Small Pox is a Judgement of God, sent to punish and humble us for our Sins; and what shall we so evade it, and think to turn it away from us? . . . This Method tends to take off the fears of this Distemper from the Minds of the People; and who knows of what Spiritual Advantage these fears might be to them? . . . God has predetermin'd and fixed the Period of every ones life . . . so that if this time be come Inoculation will not save the Persons Life."

Closer to our times, there is the example of opposition to blood transfusion on the basis of the Bible's admonition that the soul resides in the blood. Similar objections have been voiced to heart transplantation for fear that the personality of the dead donor would live on in the recipient. In the 1970s recombinant DNA technology was considered an enemy of mankind because of the power of life it transferred into the hands of mankind. In vitro fertilization was bitterly fought because of its potential to belittle the sanctity of sexual intercourse. And yet, in every case, today the majority of people would agree that these are good and compassionate uses of technology. So what changed over the years? It was simply a calming of our nerves and the manner in which the technology is described to the patient.

We live at a moment when inflammatory rhetoric distorts the debate. Hard as it may sometimes be to stand up to the naysayers, those of us at the laboratory bench have to pick up the pen and with courage and determination communicate our vision of the path we believe mankind should follow.

Let us turn to the teleological issues. If we as a society can find the courage to carefully weigh the potential of the immortal cell, what good might come of it?

Here the possibilities are as expansive as the horizon. We have a technology to reverse the process of cellular aging and from the resulting totipotent cells make any kind of cell in the human body. These new young cells will be identical to the patient, so as to prevent rejection. As a consequence of being able to extend the life span of the cells, we can grow them extensively, enabling the technician to introduce any number of targeted changes in their DNA sequence. In other words, we can erase and reintroduce DNA sequences in any way we want. Just as I am at this moment rearranging words in this sentence using a word processor, so we can now find genes and replace them with those of our own design.

Put all these extraordinary developments together and one can begin to appreciate the awesome potential of regenerative medicine. I see a world where we may make new, young, healthy cells to replace the old ones in the body. In a manner not unlike the way we replace worn-out tires or brake linings on a car, I imagine medicine one day giving us back cells and tissues identical to those we had when we were born.

Making these new medical therapies a reality at this point in history is not unlike President Kennedy's charge to reach the moon in one decade. The technology was within reach when he laid this challenge before the scientists of his day, but a mountain of organized effort was needed to make it a reality. In the same way, we have the basic discoveries within our reach to put regenerative medicine into the hands of physicians. We are missing only two components—an organized effort and time.

Regenerative medicine is currently moving forward because of the labors of a few dedicated scientists. At the rate it is currently advancing, it will take more than a decade even to begin to see its fruits. But the day will dawn when there are thousands of researchers, and that will be the time when therapies will be available for you and me.

What therapies will arrive first? It is always difficult to make accurate predictions in science. Scientific research is like walking into a dark room full of furniture. Researchers sometimes bump their legs into unexpected facts, other times walk easily where they thought they could not go. Still, based on our knowledge of cell biology, I think we can make some educated guesses.

The first uses of regenerative medicine will likely be to make and replace cells, as opposed to complex organs like kidneys or lungs. When some cells become sick or die and are not replaced, their loss directly causes disease. In diabetes, for instance, the loss or dysfunction of the beta cell in the pancreas leads to the inability to make insulin, and when insulin is not made at the right times, the cells of the body cannot understand when they should drink up the sugar in the blood—whence the disease. Yes, diabetics can take insulin shots, but this is only palliative. The crude timing of insulin delivered in this way is not precise enough to prevent terrible damage to the body. As a result, people taking insulin are still at high risk of blindness or circulatory problems that can force the amputation of arms and legs. All this could change with therapeutic cloning. One can imagine the day when doctors swab the inside of the cheek of a patient to get living cells, take them back to an embryonic state by nuclear transfer, and then steer the differentiation of these primordial cells to create young, healthy beta cells that could be injected back into the body and actually cure this devastating disease.

The same goes for other diseases caused by the loss of cell function. In the case of Parkinson's, the loss of nerve cells in a part of the brain called the *substantia nigra* leads to a gradual inability to control movement. As with diabetes, physicians can prescribe drugs that improve the symptoms, but ultimately the drugs help only a little. With time, Parkinson's will lead to such stiffness of motor function that the patient will be unable to drive a car, then unable to walk, and ultimately unable to breathe. Again, therapeutic cloning may allow us to manufacture healthy new brain cells in a dish and therefore spell a cure for the afflicted.

And the list goes on and on—spinal cord injury, liver and kidney failure, skin burns, and many more afflictions caused by the loss or dysfunction of cells. It has been said that there is not an area of medicine that could not *potentially* benefit from therapeutic cloning. This may sound hyperbolic, but we have to remember that virtually any kind of cell can be made, and these can be genetically modified in any way.

The mental image that flashes before my eyes is of a "breadboard," the ceramic plate full of holes upon which electrical engineers design inventions. Imagine the day after the transistor was invented, a breadboard piled high with electrical components, transistors, capacitors, resistors, and batteries, and the engineer saying, "We can make things with this!" In the same way, the biologists in our day have before them the "components" of life, the individual genes in the DNA code now completely sequenced. With cloning it is possible to lay our hands on the thread of life, DNA. We can shape it in any way and then, through cloning, turn those cells into any other kind of cells, or into live animals, and of course people. What mankind can do with this power is as limitless as the human imagination.

But will mankind now be able to do the unthinkable; will we be able to utilize the immortal cell to impart to the body the immortality enjoyed by the germ line? Can we make people who have the potential for immortality?

Immortality means forever, and forever, by definition, can never be achieved. But I believe for the first time in history that we can talk about some concrete applications of these new strategies in interventional gerontology. As a gerontologist, I am particularly intrigued about the possibilities of making young bone marrow stem cells. These cells normally reside inside our largest

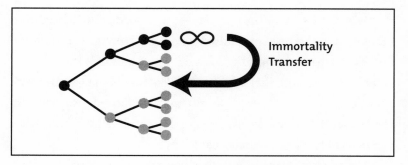

Figure 17. The Modern Scientific Version of Gilgamesh's Epic. Science has cultured the cells of the human germ line, found a means of making them identical to the patient, and identified the molecular basis of cell immortality. The remaining challenge is to identify the means of transferring the products of this technology into the body.

bones (such as the femur, the bone of the upper leg) and give rise to all of the blood cells. As we age, these cells progressively lose their telomeres and become dysfunctional. As a result, the elderly have greater difficulty mounting immune responses to the flu and other infections that would once have cost them only a day of work.

Here is where regenerative medicine could have a dramatic impact. Just as a cloned blastocyst would look the same as the blastocyst that originally led to you and me, so would the primitive bone marrow stem cells made by cloning. Gerontologists believe that the young bone marrow stem cells made by therapeutic cloning would be indistinguishable from those that you and I had when we were born. And these cells are relatively easy to transfer back into the body of an older patient. They can be simply infused into the blood vessel in the arm, and they will migrate through the blood and eventually take up residence in the bone marrow to make young blood cells instead of the old ones the patient made before. This single application of therapeutic cloning in geriatric medicine could improve the lives of millions. If so, it would be the first time in history that geriatric medicine applied scientific knowledge of the aging process in such a profound manner.

A parallel, and perhaps even more significant, application would be to make a sister cell to the blood-forming stem cell, the one called the *endothelial precursor stem cell*. Many cellular gerontologists have long suspected that the cells that line the interior of all of our blood vessels age, and when they age they can lead to the largest cause of death in the United States, which is heart disease. It is thought that as these cells age, they aggravate atherosclerosis, otherwise known as "coronary artery disease."

If, as they say, "you are only as old as your arteries," the day may come when patients will be as young as their young endothelial cells—a course of treatment that would prevent the horrendous disease that took my father from me and kills more people every year than any other.

The treatment would work like this: A cell would be taken from the patient, any living and healthy cell, and it would be taken on a ride in the time machine of nuclear transfer. This time the young embryonic cells would be differentiated into the other kind of bone marrow stem cell that gives rise to the vascular endothelium. These vascular stem cells could then be released into the blood of the patient, where they would then take up residence in the bone marrow—this time to provide a steady source of young endothelium, to reline the inner surface of the miles and miles of blood vessels in our body with young healthy cells. The impact of such an exciting new therapy could extend beyond atherosclerosis to heart failure, geriatric skin ulcers, and many other manifestations of the aging process.

Even dreaded cancer may benefit from techniques such as these. We now know that an Achilles' heel of cancer is the ability of the tumor to recruit new blood vessels to feed its campaign of unlimited and unregulated growth. The ability to genetically engineer the blood vessel–forming endothelial cells through therapeutic cloning has the potential to give tumors a "stroke"— to choke off their blood supply and to open the door to therapeutic strategies never before imagined.

The critic will be quick to pounce on these dreams and say, "Well, all of this is pure speculation—none of it has actually been demonstrated."

Yes, it is true—these are still largely the dreams of the medical research community. But the official report of the National

Academy of Sciences, a gathering of the finest scientists in the United States organized to advise Congress on matters of science and technology, stated: "In conjunction with research on stem cell biology and the development of potential stem cell therapies, research on approaches that prevent immune rejection of stem cells and stem cell–derived tissues should be actively pursued. These scientific efforts include the use of a number of techniques to manipulate the genetic makeup of stem cells, including somatic cell nuclear transfer."

We all agree that these are *potential* therapies. But the conclusion of the scientific community is that banning them before they can prove their worth would be aborting one of the most promising new technologies in medical research. The teleological argument that the ends justify the means requires that the medicine be allowed the time to give birth to lifesaving technologies.

Still, there is a nagging anxiety in all of us that such an optimistic vision cannot altogether conceal. I'm referring to the nightmare that haunted my generation of technology run amok. When brilliant and sensitive human beings gathered in the 1940s to unlock the power of the atom in the project to build a nuclear weapon, it is certain that at one point they paused at the blackboard, chalk in hand, and shivered at the thought of the numbers etched on the board. That thought must have returned to them in troubled sleep after the power they had unleashed was manifested in Hiroshima and Nagasaki.

In the same way, I cannot dismiss the dark side of this technology. For the first time in our history, we not only have the ability to make cells genetically engineered to cure disease, we can make people engineered by the hand of man. That is to say, today we have the power to reach deep within the nucleus of the cell and replace its DNA with DNA manufactured in the laboratory, or with DNA from another species. Through cloning we could then turn these engineered cells into engineered people. One can imagine the day when we shuffle chromosomes to make children with genes contributed by multiple parents and designed to enhance human traits beyond anything we could call normal. The possibilities of what mankind can make of the germ line are infinite, and the temptation to accelerate human evolution from its historic snail's pace is great. We are going to be able, in sum, to "play God."

I don't lose sleep about scientists who would introduce genetic modifications into somatic cells for medical therapies—what we call *somatic cell genetic modification*. My nightmare is that the hubris of some scientists would have us engineering *people* from genetically modified cells. Their goal is "enhancement," to make "superpeople," individuals better than any living on the planet today. When these clones, in turn, had children, they would then spread these man-made genetic modifications into the human gene pool, a process called *germ-line genetic modification*. Here is where (in my opinion) we could go terribly, terribly wrong. Since we are trying not to cure disease but rather enhance healthy life, our standards of safety should be so high that I doubt we could ever meet them.

With regenerative medicine, as with every new technology all the way back to the discovery of fire, there is risk and danger for humankind, just as there is incredible promise. Does that mean we should shy away from our responsibilities? Are we to cope with the future by hiding from it? That has not been the path we have chosen before. When it was said that mankind had no right to travel to the moon, that God would strike us down as he did the Tower of Babel, we responded courageously, convinced that despite our irrational fears, it was the good and right thing to do.

Because the application of technologies derived from the immortal substratum of life is artificial and man-made, some voice the criticism that it is *contra naturam*—that is to say, against nature. In contrast to the older concept of *contra deum*—that is, that a given act would provoke the wrath of the gods—the concern that our acts are against nature is as concrete as biology.

A champion of the "acts against nature" argument is Francis Fukayama, who charges in his book *Our Posthuman Future* that modern biotechnology is threatening the fabric of humanness itself. A member of the President's Council on Bioethics, Fukayama states, "While it is legitimate to worry about unintended consequences and unforeseen costs, the deepest fear that people express about technology is not a utilitarian one at all. It is rather a fear that, in the end, biotechnology will cause us in some way to lose our humanity—that is, some essential quality that has always underpinned our sense of who we are and where we are going, despite all of the evident changes that have taken place in the human condition during the course of history. Worse yet, we might make this change without recognizing that we had lost something of great value. We might thus emerge on the other side of a great divide between human and posthuman history and not even see that the watershed had been breached because we lost sight of what the essence was."

It would be right for Fukayama to point out that I am advocating the view that humans are machines, in reality reproductive machines, designed to pass on the reproductive germ-line cells and then to be cast off like an old coat. Fukayama is saying that the egg is a chicken's way of making another chicken. I am saying that the chicken is an egg's way of making another egg. This is indeed a different view of the meaning of human life and our essence.

Much of what we commonly call the meaning of life resides within this question. If you believe that an egg is a chicken's way of making another chicken, you side with the majority of people. Such a view of life, based on creation, is consistent with the belief that we were created specially by the creator as adults, with adult minds and bodies, and the immortal sperm and egg cells were created as a means of reproducing ourselves.

The view that the soma is the egg's way of making another egg is the evolutionary view. Not only does the theory of evolution deflate our view of the individual's significance in the biosphere, it adds insult to injury by implying that we are not even important enough to keep around after we have reproduced ourselves and brought up our children to do the same. We are designed to be a disposable soma. As William Butler Yeats put it, "Old age is nothing but a tattered coat upon a stick."

The creationist perspective is more attractive than the evolutionary one. The adult animal certainly seems to be more important. It is the chicken that undergoes the wonderful metamorphosis from a single cell to an animal that breathes, possesses a beating heart and a brain, and can learn, rear a family, and spend its days pecking in the summer sun. The egg, however, is an innocuous-looking thing. It just sits there, apparently doing nothing, until it becomes another chicken.

From the germ-line perspective, though, a person who walks through a spring forest enjoying the animals, insects, and flowers is in fact seeing an almost illusory view of life. He or she is following the somatic perspective, seeing the mortal soma rather than the immortal germ-line cells, the reproductive machines that generate the true immortal life, the germ line. The person who sees the world from the germ-line perspective instead sees egg cells using robins to make more egg cells, and pine tree germ cells using pine trees, pine needles, and pinecones to reproduce themselves. And of course, in the case of our own species, this person sees human germ-line cells using humans to gather the building blocks of life, and reproduce themselves. In essence, the germ-line perspective on life sees the body as dandelion fluff that, though beautiful, is merely a convenient but transient transport vehicle for the precious and immortal seed.

As an individual, I find the germ-line perspective of life distasteful, and I can see why Fukuyama sees it so. It *does* feel posthuman. But I value truth more than preference. Once we acknowledge the facts of science, we can begin to build a new and better world that we *do* like—a more *human* world. If we really do love our fellow human beings, then we can use this knowledge to storm the throne of the immortal germ line and usurp its power over life and death—a frontal attack on the life cycle. This is the promise of the green face of Osiris, it is the eating of the bread crushed from the body of Demeter, it is the blood released from the grapes of Dionysus, it was the green extract in my beaker, it was Gilgamesh's quest to get ahold of the green extract of immortal life.

I see Fukayama's critique as a ghost story intended to generate fear, and like a ghost story, its antagonist doesn't really exist. When regenerative medicine is applied in practice, it is by design a balm to alleviate the suffering of our fellow human beings, not to create some kind of brave new world from which we can never escape. A realistic appraisal of the realities of our evolutionary origin need not make us less human than we truly are. I would turn Fukayama's argument on its tail and suggest that the challenges we face in treating a Niagara Falls of health-care expenditures and the pressing demographics of the aging post–World War II baby-boom generation are more fearsome than any mythical monster created by a mad scientist. If we turn a deaf ear to this flood of human misery, then, yes, we as a society may lose something precious and delicate in our collective psyche. We may lose our tenderness, our compassion, our love for our fellow human being. We may reflect the lack of importance nature places on the individual. We may become as indifferent as nature itself.

In 2002, I participated in a panel discussion held at the Royal Society in London to discuss the ethical implications of embryonic stem cell and cloning technology. One of the members of the Society asked our panel to consider a dying child in need of a bone marrow transplant. Among the choices before us were adult stem cells that might be rejected and cause the death of the child or the use of therapeutic cloning that would result in the death of a preimplantation embryo. "Which death do you hate more?" I asked.

The first doctor to respond, representing the Catholic perspective, dodged the issue entirely, at which point—protocol or not—I leaned into the microphone and challenged him, saying, "You didn't answer the question!"

What he said next simply stunned me. I don't think I could have been more horrified if I had witnessed a murder. He answered: "Ultimately, I think because we are here for an eternity and our life is a small thing, I would actually choose the option of death for the child, however tragic that may be as an individual or as a parent. Because ultimately we are what we do."

There was a dead silence in the auditorium. The doctor believed in the immortality of the soul, and so, in his mind, our life spent on earth is a "small thing" compared to eternity. And because in the Christian tradition, the morality of our actions determines whether we will reside for an infinity in heaven or hell, ultimately, "we are what we do." This doctor was willing to vacate his natural affection for the child in exchange for a thinly veiled theological point of doctrine.

I am not willing to do so. A beautiful child dying on an operating table, hoping for the opportunity to see the world and enjoy the things we all take for granted, would have that opportunity denied by the Jehovah's Witness who believes a blood transfusion is against the Bible and the will of God. And on that point, the Jehovah's Witness has solid ground to stand. For the God of the Old Testament forbids the eating of blood on the basis that the soul is in the blood. It is certainly logical that God would therefore not want you to inject it into your body! Nevertheless, out of respect for the individual human life, we as a society overrule such conflicts, despite our tolerance for varied religious belief. We simply don't allow families to deny medical care on that basis.

Some members of the religious establishment have in turn attacked scientists such as me as the authors of the "Culture of Death." I see it from the opposite perspective—those who would deny a dying patient lifesaving therapy on the basis of obscure theological speculation are the culture who promote death, however unintentionally. Many thousands of innocents have suffered as a result of arguments about the divine nature of the Trinity, or the means of applying water in baptism. To my mind, the theological dispute is a small thing compared to the infinite value of the individual human life.

As we enter the third millennium, I believe that our guiding light should be compassion for our fellow human being. The capacity human beings have to love one another is among the most valued of all human characteristics. The myths of Gilgamesh and Isis and Osiris, indeed much of the folklore we call mystery religion, were all stories of the battle of love over death. A dozen books would not suffice to describe the broad and deep ocean of similar beliefs held by humans throughout history. The struggle is archetypal. Our capacity to love one another, not merely in word or pathos, but in deed and in reality, sums up all that is noble in man.

Mankind has always been at war with death. It is in our very nature to be this way. It is who we were, are, and will become. The thirst for eternity is the manifestation of human love. As the Spanish philosopher Unamuno put it, "The thirst of eternity is what is called love among men, and whosoever loves another wishes to eternalize himself in him. Nothing is real that is not eternal." In these opening years of the third millennium, the human species will inevitably begin to harness the power of our technology in the interests of that love. Immortality itself is not achievable. Whether it occurs by accident or by the final extinction of the universe itself, death will eventually prevail. But as human beings, our duty and our destiny is, and always will be, to strive for the life of those we love. Like Gilgamesh, our quest will always be to seek and find the means of bringing to our people the green essence of *zoë*, the plant that bestows the immortal renewal of life, the plant the ancient Babylonians called "*shibu issahir amelu*," that is, "The old man becometh young."

Further Reading

Angus, S. *The Mystery Religions and Immortality.* New York: Carol Publishing Group, 1966.

Budge, E.A. Wallis. *Osiris and the Egyptian Resurrection.* 2 Vols. New York: Dover Publications, 1973.

Comfort, A. *The Biology of Senenscence,* 3rd ed. New York: Elsevier North Holland, 1979.

Frazer, J.G. *Adonis Attis Osiris: Studies in the History of Oriental Religion.* New York: Macmillan, 1935.

Green, R.G. *The Human Embryo Research Debates.* New York: Oxford, 2001.

Hamilton, D. *The Monkey Gland Affair.* London: Chatto & Windus, 1986.

Hayflick, L. *How and Why We Age.* New York: Ballantine Books, 1994.

Lamont, C. *The Illusion of Immortality.* London: Watts & Co., 1952.

Larsen, W.J. *Human Embryology.* New York: Churchill Livingstone, 1993.

McGrady, P.M. *The Youth Doctors.* New York, Coward-McCann, 1968.

Meyer, M.W. *The Ancient Mysteries: A Sourcebook.* Philadelphia: University of Pennsylvania Press, 1987.

Rosenfeld, A. *Prolongevity.* New York: Alfred A. Knopf, 1976.

Silver, L.M. *Remaking Eden.* London: Weidenfeld & Nicolson, 1998.

Wilmut, I., and K.H.M.S. Campbell. *The Second Creation.* New York: Farrar, Straus, & Giroux, 2000.

Index

Page numbers of illustrations appear in italics.